005.25   Heckman, Rocky
HEC

Designing platform independent
mobile apps and services

DESIGNING PLATFORM INDEPENDENT MOBILE APPS AND SERVICES

IEEE PRESS

About IEEE Computer Society

IEEE Computer Society is the world's leading computing membership organization and the trusted information and career-development source for a global workforce of technology leaders including: professors, researchers, software engineers, IT professionals, employers, and students. The unmatched source for technology information, inspiration, and collaboration, the IEEE Computer Society is the source that computing professionals trust to provide high-quality, state-of-the-art information on an on-demand basis. The Computer Society provides a wide range of forums for top minds to come together, including technical conferences, publications, and a comprehensive digital library, unique training webinars, professional training, and the TechLeader Training Partner Program to help organizations increase their staff's technical knowledge and expertise, as well as the personalized information tool myComputer. To find out more about the community for technology leaders, visit http://www.computer.org.

IEEE/Wiley Partnership

The IEEE Computer Society and Wiley partnership allows the CS Press authored book program to produce a number of exciting new titles in areas of computer science, computing, and networking with a special focus on software engineering. IEEE Computer Society members continue to receive a 15% discount on these titles when purchased through Wiley or at wiley.com/ieeecs.

To submit questions about the program or send proposals, please contact Mary Hatcher, Editor, Wiley-IEEE Press: Email: mhatcher@wiley.com, Telephone: 201-748-6903, John Wiley & Sons, Inc., 111 River Street, MS 8-01, Hoboken, NJ 07030-5774.

DESIGNING PLATFORM INDEPENDENT MOBILE APPS AND SERVICES

ROCKY HECKMAN

WILEY

Library of Congress Cataloging-in-Publication Data:

Names: Heckman, Rocky, author.
Title: Designing platform independent mobile apps and services / Rocky
   Heckman.
Description: Hoboken, New Jersey : John Wiley & Sons, Inc., [2016] I Includes
   index.
Identifiers: LCCN 2016009419| ISBN 9781119060147 (cloth) I ISBN 9781119060185
   (epub) I ISBN 9781119060154 (Adobe PDF)
Subjects: LCSH: Mobile computing. I Cell phones–Programming. I Mobile apps.
Classification: LCC QA76.59 .H43 2016 I DDC 005.25–dc23 LC record available at
   https://lccn.loc.gov/2016009419

Printed in the United States of America.

10 9 8 7 6 5 4 3 2 1

Thank you to all my friends who finally convinced me to write a book. Most of all, thank you to my wife Stefanie, and my two beautiful girls Elyssia and Keiralli for not only pushing me to finish, but putting up with all the time I spent doing it. I love you all very much.
Look Dad, I did it!

TABLE OF CONTENTS

LIST OF FIGURES

LIST OF TABLES

PREFACE

I WAS WONDERING how to start this preface, and it occurred to me that writing one is backwards. After all, this is the text before the book, but you write it after the book is finished. I suppose that is analogous to how we have written software for a long time. We figure out what it is we are trying to do usually though creating systems that do what we think we want them to do; then we go back and write the documentation about what the system actually does.

This book, in large part, is aimed at helping to make that a more harmonious effort. Technology is moving so fast now that often we find ourselves trying to create mobile apps and services in a very reactionary manner. We tend to be on the back foot and playing catch up most of the time. If we could just take the time to sharpen the proverbial axe, we'd be able to get more accomplished, faster, with a lot less hair pulling.

I suppose over the past couple decades of doing this, I've seen that pattern time and time again. But there are also some good habits and patterns that I've seen along the way that in some respect were ahead of their time. Service-oriented architecture, for example, was a great idea for connecting the myriad of systems we've had inside organizations with simple, easy-to-use interfaces. In fact, this tried-and-true pattern, or collection of patterns if you will, is more relevant today in our commodity cloud computing, mobile app world than ever before.

This book is designed to provide some high level architectural guidance on how to design modern mobile apps and services so that they can adapt as technology inevitably changes. Of course, we start off with a brief history of our mobile computing explosion, and take a look at attempts to create cross platform apps and technology stacks.

Then I want to introduce what hopefully has become an obvious application stack. While we have been fairly fixated on a *N*-tier stack, where *N* usually equaled three, to truly futureproof our architectures, we really need two more clearly defined layers to provide us an abstraction boundary which insulates our code from changes in external client technology, as well as the rapidly changing data storage technologies we use today.

Once we have our layers sorted out, we'll have a look at various patterns of application development and how they apply to this layering system to create performant and resilient services for making powerful mobile applications.

I hope that you find this guidance useful. Perhaps it will make you think of things in ways you hadn't before, or validate thinking you've already implemented. In any case, I hope it prevents you from having to operate in a reactionary manner to the rapid changes of our modern computing world and lets you get on the front foot so you can focus on creating great apps and services instead of retooling everything because a new phone hit the market.

Target Audience

This book is for anyone who is responsible for the design, architecture, and development of modern mobile apps and the services that support them. I've written this book with futureproofing in mind. Ideally, the architectures and patterns in this book will provide you with an approach that will futureproof your designs.

By following this guidance, you should be able to create mobile app services that you can adapt, modify, update, change, or integrate without disrupting your mobile apps, or your teams. You should be able to deploy new services, change existing services, and add new client apps all without disturbing any of the running systems.

Most of all, you should be able to adapt services and apps based on this guidance to any new mobile platform that comes along. This will greatly increase your code reuse, make your teams much more efficient, and make your organization adaptable to the ever-changing mobile app landscape.

ACKNOWLEDGMENTS

These kinds of things don't happen without a lot of people in the background pushing, pulling, helping, and sometimes simply nodding and smiling. I would like to thank Chris Bright for encouraging me and allowing me the time to put this together. I'd also like to thank Andrew Coates and Dave Glover for letting me harass them with ideas, and "what if" questions all the time.

Most important of all, I need to thank my wife Stefanie Heckman, and my two girls Elyssia and Keiralli. They not only encouraged me to finish, but were patient with me, and gave me the time to keep typing away. I think, in the end, their love and enthusiasm are what really got this book over the line. So if you like it, don't forget to thank them too.

CHAPTER 1

THE MOBILE LANDSCAPE

1.1 INTRODUCTION

When the idea of reaching people first struck home in the dark ages, we wanted to find ways for people to use and pay for our services. We had to find a way to let people know these services existed. In early days there were town criers, then during the industrial revolution when we could reproduce and distribute text to a largely literate audience, we had broadsheets. Then came the catalogue where mercantile companies would list their wares for sale. Once we heard the first radio waves, one of the first things they did was to sell advertising on the radio. This graduated to television advertising. Then along came the Internet. Everyone had to get themselves a website and would put their website address in their print, radio, and TV ads. Along came Facebook and everyone created a Facebook page for their companies.

Now, everyone wants to have an app for their customers to download. These apps go with customers wherever they are and provide instant interaction between consumer and supplier. We can push advertising into them, take orders through them, keep in touch with friends and relatives, and of course play games, listen to music and watch videos all in the palms of our hands. These experiences require devices, operating systems, and apps, all of which require software companies, architects, and developers to produce them. Unfortunately, these devices and operating systems often change.

In today's computing world, there is one thing you can be sure of; the leading operating system (OS) platform will change. As recently as 7 years ago, Microsoft Windows Mobile was the leading smartphone platform and tablet computers, while mobile, were large and clunky and ran full versions of the Windows XP and Windows 7 OS. Then came Blackberry which took a lot of market share from Windows Mobile. But that only lasted until the iPhone came along in 2007 and we went from a feature phone dominated world to a smartphone dominated one. This set a new benchmark and became the leading mobile computing platform. In the same year, the Open Handset Alliance re-released Linux-based Android-powered smartphones. Then in 2010, Google launched its Android-based Nexus series of devices. By 2011, Android-powered smartphones made up the majority of mobile OS-powered smartphones shooting past the iPhone.

While phones were taking off, in 2010 Apple released the iPad. Tablet computing was not new and in fact Microsoft and its Original Equipment Manufacturers

Designing Platform Independent Mobile Apps and Services, First Edition. Rocky Heckman.
© 2016 the IEEE Computer Society, Inc. Published 2016 by John Wiley & Sons, Inc.

OEM partners had been trying to sell tablet computers since 2003. However, the iPad's sleek design brought tablet computing to the masses despite the clumsy and restrictive iOS operating system. This opened up the tablet computing market which Android was well suited for. After the iPad's initial success, by 2013 Google's Android-powered tablets had overtaken iPads as the tablet of choice. Additionally, although lagging considerably behind, Microsoft has re-created itself to make a run in the mobile and tablet computer markets as well. With Microsoft's massive install base, and very large developer ecosystem, they are likely to challenge Apple and Google in the mobile and tablet space eventually. With Windows 10 released as a free upgrade for over 1.5 billion eligible devices [1], it is likely to be the most common cross platform OS. That is, over half of Gartner's predicted 2,419,864,000 devices shipped into the market in 2014 [2]. Overnight the app ecosystem market leader could change again.

What this means for software developers, independent software vendors (ISVs), hobbyist app developers, and online service providers is that every few years they will have to retarget their efforts for a new platform, new development languages, new development tools, new skills, and new ways of thinking. This is not an attractive proposition for anyone. However, due to the success of the iPhone and iPad, software developers were willing to re-skill and even purchase proprietary hardware just to be able to develop applications for the new platforms. Then when Android devices surpassed the Apple devices, these same developers painfully went through the whole process again. Developers were forced to maintain three or more separate and complete codebases. This is the problem that Platform-Independent Delivery of Mobile Apps and Services solves.

If you are not planning a platform-independent strategy, you will likely be an ex-company in 3–5 years. Due to the rapid change of the consumer and enterprise mobile computing landscape, software developers must be able to adapt to new platforms, devices, and services before their competition. While cross-platform goes a long way toward this goal, it is still cumbersome and tends to lag behind a more platform-independent strategy. While it is not practical to get completely away from device-specific app code, the more you can move off of the device and put into a reusable back-end service, the less code you have to write and maintain when a new OS version or a new platform comes along. In this book we will examine strategies to do this, and provide future proof foundations to support changes in the computing landscape down the road.

Disclaimer: This book was written in late 2014. Everything in it was accurate at that time. If you are reading this in 2025, expect that a few things have changed. Just keep this in mind as we go through this so I don't have to keep writing "At the time of this writing…."

1.2 PREVIOUS ATTEMPTS AT CROSS-PLATFORM

1.2.1 Java

"Write once, run everywhere" was a slogan developed by Sun Microsystems which promised cross-platform capability for Java applications supported by Duke, Java's Mascot shown in Figure 1.1. This gained significant traction in the mid to late 1990s.

Figure 1.1   Duke

In the beginning of this era, the promise seemed legitimate. You could write the Java code once, package up your Java byte code in .jar files and run them on any system that had the Java Virtual Machine (JVM) interpreter. It worked so well that there are even C to Java compilers so your C applications can run with the same cross-platform reach that Java had.

The promise was that you could write code like this:

```
class CrossPlatformApp {
    public static void main(String[] args) {
        System.out.println("I am omnipresent!");
        // Display the string.
    }
}
```

And it would run on every computer and device that ran Java without compiling multiple versions for each target device. All you had to do was make sure that the target device had the correct version of the JVM installed on it.

This worked fine until various vendors started creating their own versions of the JVM to run on their platforms. By 2014, more than 70 different JVMs [3] had been created that could run Java applications, for the most part. The catch was that they were each slightly different.

If we take the Sun JVM to be the standard, some of these other JVMs were better, and most were worse, at interpreting Java byte code. Some of them such as the IBM J9 (http://en.wikipedia.org/wiki/IBM_J9), the Azul Zing JVM (http://en.wikipedia.org/wiki/Azul_Systems), and the Microsoft JVM (http://en.wikipedia.org/wiki/Microsoft_Java_Virtual_Machine) were better and faster than the original Sun JVM. They even went so far as to add extra features and some constructs that were more familiar to traditional C/C++ programmers in order to make the transition easier for them.

While this seemed fantastic at the time, because it meant every platform vendor had a JVM to run Java, they weren't all the same. So what may work on the Sun JVM may not work on the Microsoft or IBM implementation. Even though some of these implementations such as the 1998–1999 Microsoft JVM outperformed the Sun version, they weren't entirely compatible with the Java 1.1 standard. This lead to Sun suing Microsoft and other JVM vendors in an attempt to try to defragment the Java playing field. The result was these other vendors stopped supporting their proprietary versions of the JVM and true high-performance, cross-platform capability for Java applications started to deteriorate.

This is a trade-off that you see repeatedly in cross-platform development. There has always been a compromise between running on many different devices, and getting as close to the hardware as possible for fast execution. It's the nature of computers. Each device may have slightly different hardware running the code. This means that the operating system and CPU may understand different instructions on each device. Java tried to solve this with the JVM. Different JVMs are written for the different environments, and they provide an abstraction layer between your Java code, and the nuances of the underlying hardware. The problems arise when one JVM interpreted the incoming Java code slightly differently than the next one and the Java dream becomes fragmented.

While Java is still widely used for applications, there are many versions of it depending on what kind of applications you are writing. There are four primary versions of Java that are supported by Sun.

- Java Card for smartcards.
- Java Platform, Micro Edition (Java ME) for low powered devices
- Java Platform, Standard Edition (Java SE) for standard desktop environments
- Java Platform, Enterprise Edition (Java EE) for large-scale distributed internal enterprise applications

All of them require a very standards adherent JVM to be installed on the target machine for them to run. Often the JVM can be packaged up with the application deployment, but the dependence on the JVM and specific versions of the JVM have made cross-platform Java apps troublesome. This is largely because you can never be sure of the JVM on the target device.

This is a common issue with most interpreted languages such as Java, Python, Ruby, .NET and any other language that is Just-In-Time compiled and run in a virtual environment or through a code interpreter. These kinds of things also reduce the speed of the applications because everything is interpreted on the fly and then translated for the CPU rather than being compiled down into Assembly or CPU level instructions which are executed by the CPU natively. This is why C and C++ and similar languages are referred to as native languages.

So while Java was a very good attempt at write once run anywhere, it fell short due to its dependency on the JVM. It still has a large install base and works very well in many web app scenarios. It is also the primary app development language for Android-based devices which at the time of this writing was the world's leading mobile device operating system. Java can also be used to create apps for Apple's iOS-based devices through systems such as the Oracle ADF Mobile Solution [4, 5]. However, the vast majority of iOS targeted apps are written in Objective-C using Apple's Xcode environment. Due to the difficulty of developing with Objective-C, Apple introduced a new language called Swift for the iOS platform to help combat the hard translation form Java or C# for iOS developers and to improve the performance over Objective-C apps. At the time of this writing, Java did not work on the Windows Modern apps or Windows Phone platforms. Java does still work in the Windows Pro full x86 environment.

Java is perhaps the closest the industry has come to write once run anywhere. But it has been plagued by spotty JVM support, and a push toward more proprietary development for iOS and Windows to get the speed and integration to a more seamless state.

1.2.2 Early Web Apps

On August 6, 1991 the first website was created by Tim Berners-Lee at CERN (http://info.cern.ch/hypertext/WWW/TheProject.html). Ever since then, we've been pushing web browsers beyond their intended limits. From their humble beginnings as static pages of information to the preeminent source of all information and social interaction, websites and web apps have become ubiquitous in our connected world. It was inevitable that web access from every computer would lead to web apps being seen as the next great cross-platform play.

Web apps have been a popular attempt to run anywhere. All you need is a web browser on whatever device you have and you can use web apps. Well, that's the idea anyway. In reality this proved much more difficult than anyone hoped. Prior to HTML 5 you naturally had HTML 4. HTML 4 was still largely just a markup language designed to handle content formatting. Web pages displayed in the browsers were largely static text and images. Then early browsers such as Netscape and Internet Explorer 3 incorporated a JVM to interpret the Java code in the web pages. Web server software such as Apache and Internet Information Services could also run server-side Java code and send the product of the code back to the browser as an HTML page.

This worked pretty well, up until Sun sued Microsoft and they stopped including the Microsoft JVM with Internet Explorer. Since at the time it had become the world's most popular browser, that was a problem for Java-based web apps. It forced users to manually install a JVM from Sun which was an extra step most people weren't overly fond of.

This resulted in some interesting changes. Netscape produced its own web programming language called LiveScript in 1995, which it then changed the name to JavaScript when it introduced Java support in Netscape 2 in 1996. Meanwhile in the same year Microsoft produced Active Server Pages (ASP) and in an attempt to get around the Java JVM problem, it also included VBScript for the coding portion in ASP. JavaScript pretty much won the client-side scripting battle when Microsoft included support for it in Internet Explorer 3 but had to call its version JScript.

JavaScript became adopted as a standard known as ECMAScript which is in its fifth edition (5.1) released in June 2011.

In order to do interesting things with web apps, we needed to do things that HTML 4 simply couldn't do on its own. So one of the first things that was built into web browsers was a JavaScript interpreter. Now you could run scripts in web pages that could do things like display today's date, manipulate text in text boxes, and rotate pictures. This was nice, but in the days of Mosaic/Mozilla, Netscape, and Internet Explorer 3, it was really pushing the envelope.

To get a bit more out of the web apps, people started developing plugins for web browsers for things like audio and video. Macromedia introduced Flash in 1996 and it opened up all kinds of new opportunities to do very advanced graphics in a web browser through the Adobe Flash Player. By 2000, Flash was everywhere and even used to produce some animated TV commercials and 2D programs [6]. Around the same time in 2007, Microsoft introduced Silverlight which was a competing technology to Flash and offered audio, video, and graphics for web apps.

At the time, HTML had been reduced down to something like the following:

```
<HTML>
<HEAD></HEAD>
<BODY>
...
</BODY>
</HTML>
```

Everything between the <BODY>...</BODY> tags were references to JavaScript and plugins of some sort that offered extended capabilities that were not part of the HTML4 specification. This included embedded audio, video, and pluggable content.

The core of the problem was that while HTML was an open standard that everyone understood and agreed on, things like Silverlight and Flash were not. This lead to controversy about its use and widespread adoption due to the dependency on proprietary technologies. In fact David Meyer quoted Tristan Nitot of Mozilla as saying:

"You're producing content for your users and there's someone in the middle deciding whether users should see your content," [7]

This sentiment essentially created a mistrust of proprietary technologies that started developers looking for standards bodies to create web standards that could fill the voids that things like Silverlight and Flash handled.

Although these technologies are still prevalent today, they met with some resistance in the mobile computing era. Some of it was due to Apple initially not allowing Flash to operate on its iOS platform. While this was fixed by allowing Adobe AIR apps to run on an iPhone which wrapped Flash content, it was enough for Adobe to re-evaluate its position on Flash and to withdraw support for Flash on mobile devices in 2011 unless it is embedded in Adobe AIR applications. Instead, they plan to "aggressively contribute to HTML5." [8]

Android having become the most popular platform by 2014 allows users to develop their apps in Java. This has brought a resurgence of development into the Java camps and solidified the once flagging language as a pre-eminent language in the mobile app development space. Eclipse tends to be the Java development environment of choice for Android apps as well.

This was an excellent move on Google's part because there were a lot of Java developers out there already. They could bring their existing skill set to developing for Android. This made the transition easy and afforded Google a huge advantage in catching up and surpassing Apple in the numbers of apps available in the Google Play store. You can also write apps for Android in C and C++ if you chose. This freedom has led to Android being the first platform people tend to release apps for.

Microsoft tried to keep Silverlight alive by allowing developers to create Windows Phone 7 apps in Silverlight. But their preference was for HTML5-based apps for Windows Phone and in versions of Windows Phone after Windows phone 8, Silverlight is no longer the platform of choice. You might say Silverlight is not a forward looking option for Microsoft. Microsoft also has a strong support for HTML5 app development in its platforms and has recently adopted the Apache Cordova framework for developing cross-platform apps in its Visual Studio product.

This myriad of technologies, plugins, mixed standards, and a very high dependence on browser version caused the web apps explosion to become muddled and difficult to develop for. Many lines of web app code are dedicated to just figuring out which browser and what version you are running in so you can do things slightly differently. You would have to check to see if a particular plugin was installed like Flash or Silverlight. You had to check to which browser version to know what kinds of scripting the browser supported and then run the version of the script for that particular browser version. To put it mildly, it was a nightmare. Having to write several different versions of the same code in the same web page to accommodate all the possible browsers their versions and plugins drove most web developers to liver failure due to excessive drinking. There were simply too many competing technologies amidst the browser wars.

1.2.3 Multiple Codebases

When it came down to it, developers knew they needed to support multiple platforms one way or another. They had executives wanting the apps on the latest buzzword compliant device. They had IT staff wanting their monitoring apps on the myriad of different devices they used and then they had their customers wanting the apps on every PC, smartphone, and tablet ever made in the 1990's. They had to come up with something so they ordered pizza and got down to duplicating their apps on every platform that 90% of their users used. Typically this meant Windows for desktops and laptops, iPhones and iPads running iOS, and Android-based tablets and phones. At a minimum, it was three separate development exercises. Windows apps were written with the usual Visual Studio and .NET combination. iOS apps required Xcode running on Apple hardware to write Objective-C apps. Android's Java-based apps were normally written in Eclipse on Windows- or Linux-based machines.

If you want to cover these three platforms you need at least two development machines: an Apple Mac and a Windows- or Linux-based machine. This was a real problem because not only did you pay more for extra hardware, but you needed more desk space. Early on, you couldn't write apps in Visual Studio for an iPhone or Android. You couldn't write Windows apps with Xcode. So you were forced to duplicate your equipment, effort, time, and skillset to cover the cross-platform world. But when you were done, you had apps on all the platforms that worked really well on the platform. Of course, if you wanted to add or change any features, you had to do it three times.

This also meant that you had to keep track of three sets of source code, and more importantly track three separate lists of bugs. If a severe bug showed up on one platform and not on the other two, you had to update and release just that one app. This could become a real problem. For example, if your Android app was full of security holes and issues that were present on some versions of Android and not others, it was very difficult to keep up with it all and keep the Android app patched without getting too far out of sync with the Windows and iOS versions.

Then of course there is the problem of the three different environments having very different capabilities. For example, you can create complex large databases for your app on Windows running on a laptop, but you can't do that in an iOS or Android app because they don't run on devices with large hard drives and lots of memory. So the Windows version tended to get ahead of the iOS and Android versions when it came to features. This often lead to developers creating their main apps on Windows, and having companion apps on iOS and Android that provided a subset of the features of the Windows version, or that complemented it extending the app to mobile devices.

As time went on and mobile computing became the norm rather than the exception, the iOS/Mac combinations were developed first, and web apps developed for internal usage or other platforms. In most cases, line of business apps is being developed for iPads, as much as for desktop computers. Server-side services are almost exclusively in WS* or RESTful web services. This has put web technology development in the lead, followed closely by native Android and iOS apps.

This is a very expensive way to do app development. It's not just the cost of the extra hardware. You duplicate your source code storage, management requirements, and man hours and worst of all; it requires some very different skill sets, which means you essentially maintain three development teams for one product. Then inevitably when the next big thing comes along, and the PC or device lead changes hands again, you have to scramble to re-skill for the new platform and create yet another set of source code or development systems.

This tends to cripple most small shops and they are forced to focus on one platform. Large development shops and large organizations with their own IT departments and development staff tend to choke on budgetary issues and time to market issues due to the complexity of launching on multiple platforms. Up until now this has made it near impossible for all but the largest companies to do multi-codebase development.

That is what this book is all about. I will cover the improvements to cross-platform standards, improvements in cross-platform development tools, and most importantly strategies you can implement to achieve cross-platform app development with minimal overlap, duplication, reskilling, and cost.

1.3 BREADTH VERSUS DEPTH

One of the things that every app developer wants is for millions of people to use and love their apps. This means they need to be able to reach as many potential users as possible. This also applies to enterprise line of business (LOB) app developers that have to develop corporate apps that will work on any device employees have in the 'bring your own device' world. The catch is that for their app to have a rich engaging experience, it has to work on whatever device the user has at the time. Unfortunately this means that if you want a rich experience on iPhone, for example, you have to develop the app strictly to run on an iPhone. This app won't be able to run on a Windows Phone or Android-based phone. If you want to reach Android users you have the same problem. You can create an amazing app on Android, but it won't run on iPhone or Windows Phone. This is the breadth versus depth problem that app developers face when deciding which platform to target as illustrated in Figure 1.2.

In the past, developers have had to decide on the trade-off between reaching as many users as possible and the rich feature set which requires very device specific code. This has led to them picking the platform that has the most local market share that day. There are effectively three relevant platforms in the mobile space at this time: Android, iOS, and Windows. In this book, I will consider these three major platforms as our baseline.

This leaves us with a problem. If you are developing apps for mobile devices, which one do you choose? Choosing one means that in a few months or years you could be cut out of 60% or more of the addressable market. Putting all of that work into something that will be usable by less than half of the market is not very attractive.

There are some alternatives. You could make a responsive website that works on all mobile devices and adapts to different screen sizes. This gives you the broad reach across the user base but web apps tend to be very restricted to the features you can use. With technologies such as Phone Gap using HTML5, CSS3, and JavaScript as their primary programming stack, you effectively develop web apps, wrap them in a native app frame, and upload them to the app store. It's just HTML though, is that feature set rich enough? HTML5 has tried to fix that.

HTML5 has given developers the ability to create fairly rich web apps that can do most basic things native apps can do. HTML5 apps can access the location information on the device, local isolated storage, stream audio and video, and even replace Flash with the canvas elements that are available. With HTML5 it is possible to create very functional business applications such as Excel Online shown in Figure 1.3.

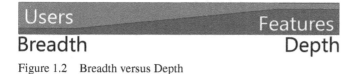

Figure 1.2   Breadth versus Depth

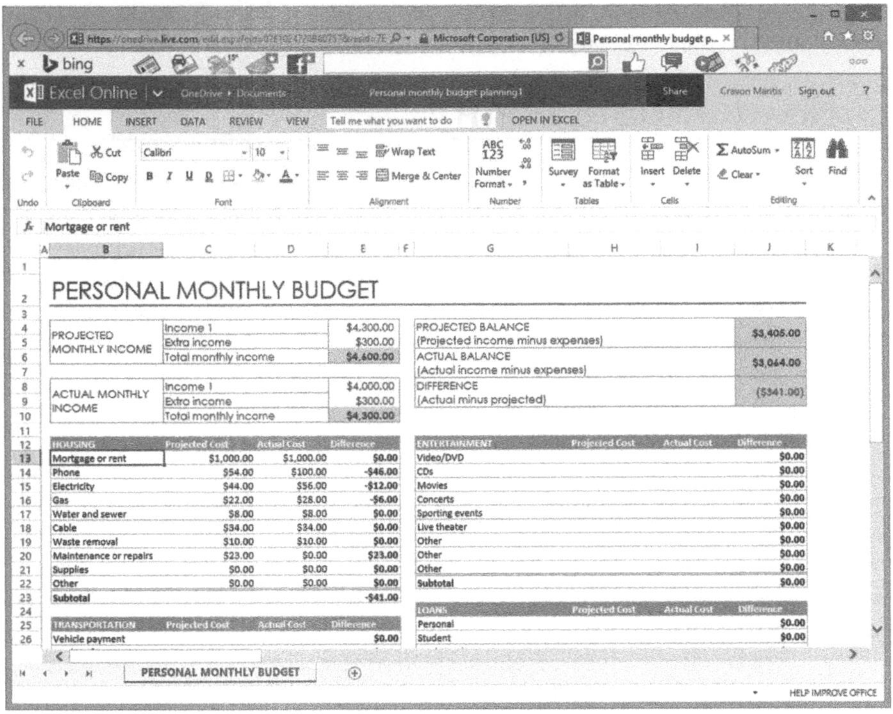

Figure 1.3   Excel Online https://office.live.com/start/excel.aspx

This is a very functional version of the Microsoft Excel that includes formatting, cell formulas, graphs, and data imports.

You can also create rich non-Flash games such as The World's Biggest Pac-Man. The World's Biggest Pac-Man (http://worldsbiggestpacman.com/) even allows users to create and upload mazes of their own for other players to play, all built with HTML5.

With HTML5 comes instant reach. Every major platform has an HTML5 capable browser that makes these web apps available to the users of those platforms. As good as this is, these web apps still lack important features like browser-independent off-line capability, access to other local documents and apps, integration with local contacts, e-mail, and calendars, etc. You still have to trade some features for reach. On individual platforms, native apps will be more attractive than web-based app in some circumstances.

1.4 THE MULTI-PLATFORM TARGETS

1.4.1 Traditional

By traditional platforms I am mainly referring to websites and clients in a client-server configuration. These can certainly be called multi-platform as different web browsers have different capabilities. Additionally clients in client-server systems can be created on any platform providing that platform will allow communications with Internet-based services. This book really isn't concerned with thick clients as they tend to be self-contained and not normally mobile. That being said, you can certainly use the information in this book to create thick client apps that talk to a service back-end for whatever business, consumer, or game purposes you want.

This segment isn't really the one that developers think of when they think of new apps. However, it shouldn't be overlooked. The three major operating systems in play at the time of this writing, where apps are concerned, are Android, iOS, and Windows. iOS runs on iPhones and iPads, but not on MacBook or Mac desktops. Android and Windows however run on many more form factors. Android runs on phones, phablets, tablets, and notebooks. As of Windows 10, Windows runs on every form factor in common use today from embedded Internet of Things devices to phones, tablets, laptops, desktops, and large screen displays plus Xbox and HoloLens (www.hololens.com).

This is also why you need to make sure you don't walk into app development, thinking you are going to create an "iPhone" app or an "Android" app. If you do, you will certainly cut yourself out of a huge share of the potential market. If you don't consider the more traditional form factors such as laptops and desktops, you will miss the majority of the addressable market. There are still, at the time of this writing, more desktop PCs and laptops in use than there are iOS devices and Android devices combined. Go into this thinking cross-platform and cross form factor from the beginning. Don't cut yourself out of the majority of the market.

1.4.2 Mobile

This is a broad category but probably the one you are interested in developing for the most. After all, this is a book about mobile apps and services. It includes pretty much everything that is computerized, that is portable, but not that you wear. Phones, tablets, phablets, laptops, and other devices like these. Any device that a user can carry likely has device location capabilities, gyroscopic sensors, and cameras which open up a lot of new scenarios and potential for your mobile apps and services. This does apply to wearable and embedded devices as well but we'll cover those in a bit.

This is also the category that most people think of when they think of mobile apps. There are a lot of new developers coming on board, probably some of you reading this right now. This is the fastest growing segment of app developers. With millions of mobile apps in various app marketplaces, and some people making billions of dollars on them, it's not very surprising that this is such a hot topic. For your app to be successful, it has to offer a compelling user experience and features. It doesn't matter if it's a game, a family management app, or a business application; it will need to stand out amongst the crowd for you to be successful.

The app explosion, and accessibility of the tools, means that anyone with a good idea and some motivation can create good apps from the comfort of their home. This segment is making millionaires by the week with apps deployed to millions of people that only cost one dollar. Why not reach as many of those potential users as you can?

To reach this kind of success, an app that is integrated with a user's lifestyle, that connects them to the world, and that offers powerful features is required. You need to add value to their device, regardless of form factor, CPU power, RAM, or battery life. This means being able to extend beyond the device for connection to the world, and connection to more computing power than is available in the users' hands. Solid back-end services will get you there. These services will provide your app an edge over the myriad of other similar apps in the marketplace that are limited to the shell of the device.

1.4.3 Wearables

This is an emerging category in the mainstream. Even though the idea of a network-connected smart watch has been around since 2003 [9] and maybe before, they are starting to get a lot of public attention and most major manufacturers are coming out with watches or some other form of Internet-connected wearable computing device. If you are a Star Trek Next Generation fan, think of the communicators built in the Insignia on the uniforms. It provided not only communications, but also vital signs reporting, location reporting, and other network-connected services. From a more mundane perspective, think of smart watches, fitness bands, and glasses.

Now we even have fully self-contained holographic displays like Microsoft's HoloLens, an entire wearable PC that is less than the size of a bicycle helmet. HoloLens enables full persistent 3D holograms projected into your world. It comes complete with everything you'd expect in a mobile device such as location, cameras, and sensors, but includes the storage and CPU power of a desktop PC without the wires and goes where the user goes. This kind of thing introduces new gesture-based computing features and possibilities that we have only just begun to imagine.

Everyone will have some kind of wearable in the near future. In some cases they may not even know it. The ubiquitous nature of wearables such as new watches that have some kind of minimal GPS capability and health monitoring in them opens up very personal computing experiences for mobile app developers.

Many wearables are touting health and medical benefits. They can track everything from duration and route of walking, bicycling, and swimming to heart rates and even ultraviolet ray exposure and potentially glucose levels [10]. These kinds of lifestyle aspects make wearables attractive to certain segments of the population. In industry, think of patients, athletes, and military personnel. The military is already using wearable computing technology to track the heath of soldiers in addition to location, and transmitting live audio and video through satellite networks to command and control centers. Professional sports are using it to track things such as how far a player runs during a soccer match, to impacts they take in NFL games which may indicate they should undergo concussion screening.

For wearables to be useful, they have to be lightweight, with extended battery life. They tend to focus on data collection and upload. More advanced ones such as the Apple Watch, Android Wear, and Microsoft Band offer limited computing capability, but tend to integrate with the user's mobile phone. This means offloading processing power is critical. Being battery friendly is just as critical. A watch that you have to charge twice a day isn't very useful. A fitness band that doesn't last a marathon is equally as useless. Glasses that make you gyrate your head and stare off into space will disrupt the normal in-public interactions. Wearables need to blend into our lifestyle and enhance it, not disrupt it. The services that power them can be powerful, and add a great deal to a user's lifestyle making them very attractive if they are built right.

1.4.4 Embedded

This category comprises all of the devices that contain an Internet-connected System on a Chip (SoC) or small computer such as Arduino, Raspberry PI or other proprietary systems. This devices are colloquially called the Internet of Things. The category isn't new, and we've had computers and applications buried in various things around the industry for many years. Cars have had embedded systems that mechanics would use to monitor and tune the car with. Racing has used this approach quite heavily for several years. Now that it is becoming easier to do with more SoC processors and operating systems, we even have drink vending machines with advanced processing capability in them [11]. These vending machines can look at you with built in cameras, and fairly accurately guess your sex, age, and height. They then feed this demographic data, along with your drink choice, up to cloud-based processing systems to aid in their marketing and machine placement choices. Not to mention the boring aspects of how much of which drink is left so the delivery person knows exactly how much of which drink to restock the machine with.

Gaining access to data that help with demographics and large-scale trend analysis is perhaps one of the largest areas of interest in the Internet of Things movement. Consider areas of home appliances. If you were a large grocery store chain owner, what could you do with a Smart Refrigerator? A fridge that knew what was in it, what the expiry dates of the various things were, and what kinds of things this particular family like to eat? With that kind of data you could automatically e-mail the family specials on items they need, recipes for food that they can make with things in the fridge that are about to expire, or allow them to integrate their food habits with their family doctor's records so he can make sure they aren't on the road to diabetes. The possibilities are boundless. It's a great time to be a mobile app service developer.

This is also the area that requires the most high performance data injection systems. When you start to consider all of the potential devices that will have Internet-connected computers in them, the numbers start to get very impressive. Consider smart utility meters. Every home will have one. How many devices are there, just in your city? How about New York, Hong Kong, or New Delhi? There are over 8.4 million people in New York City alone. If these meters and devices are all reporting daily, you will have to be able to scale to handle that traffic during reporting times, and scale back during non-reporting times or in the cases of personal devices, handle that kind of traffic on pretty much a 24/7 schedule.

So as you can see, we are in a situation in the industry where we need to think proactively about how we are going to future proof our designs. For a very long time we've been working on the back foot trying to create cross-platform technologies that haven't quite hit the mark. This is in large part due to the fact that one specific technology may never be all things to all people because by the time it's done, there will be a new platform to contend with. With proper planning and a solid architectural approach, you can minimize the impact of cross-platform development for your apps and service. In this book, I will present some of the architectural patterns and features you can use to create and deploy rich cross-platform app services that will be comparatively easy to maintain across current and new mobile platforms.

CHAPTER 2

PLATFORM-INDEPENDENT DEVELOPMENT TECHNOLOGIES

THIS BOOK IS about designing platform-independent app services. This means I'm going to keep things fairly "architectural" and component level. I'm going to try to avoid getting into too many implementation specifics such as code examples. Except for a couple of cases I'm going to try not to delve too deep into particular vendor implementations either. While this sounds disappointing to some, there's a good reason for it.

In today's mobile app services world, you can use pretty much whatever technology you want to as long as it gets the job done efficiently. There are a few technologies that are rising to the top as the broader choice in the industry, but use what works for you as long as you can build the patterns discussed in this book. If I go into too much C# and you are a Java programmer, you'll spend more time translating the code to Java in your head than understanding the bigger picture. Besides, there are thousands of books out there already on how to program in C#, Java, JavaScript, Python, Ruby, PHP, etc. They will cover the specifics a lot better than I can do in one book here. Which brings me to my Golden Rule (well for this book anyway).

The Golden Rule

I've been around a while and seen a lot of apps. I've been supporting, developing, designing, troubleshooting, securing, and consulting on software and apps for about 17 years. So I have some experience and observations from both the ivory towers and the trenches. There are a lot of books and articles out there on what's best and what you should do, but I have a golden rule. This golden rule is something you should commit to memory, don't worry it's a pretty simple rule. Throughout this book, there will be lots of advice, architectural patterns, and design guidelines that should help you create better, more efficient, and highly scalable mobile apps and services. However, you should always apply the golden rule. There are a lot of Rules

Designing Platform Independent Mobile Apps and Services, First Edition. Rocky Heckman.
© 2016 the IEEE Computer Society, Inc. Published 2016 by John Wiley & Sons, Inc.

of Thumb in this book, put those on a poster near your desk, with the golden rule at the top. The golden rule is:

Always do what makes the most sense for your situation.

I may give an example of a game that requires high-speed message transfers to hundreds of clients. Don't just say "yeah but I am writing a live trading app" and dismiss things. The principals will apply and you should adapt them for your situation. Keep your big picture in mind.

2.1 VENDOR LOCK-IN

Inevitably when I am discussing a level of platform-independent delivery or a particular cloud technology with an ISV, large multi-national, or 4 man software development shop, someone will say, "Our policy is that we do not do anything that locks us into a particular vendor." That's wonderful. Good luck with that. Now let's get to the reality of computers.

Something I want to point out, *There is no way with current technology to create a truly and utterly vendor agnostic and technology-independent system.* At some point, you are going to have to commit to using some technology. Be that Windows or Linux, Java or C#, Amazon Web Services, Microsoft Azure, or Google Cloud Services or this database or that NoSQL Store. Once you decide, there will be things that you do that conform to that technologies' manner of interacting with your developers and the greater world. Unfortunately, each technology vendor does things in a slightly different way, which means that either syntactically or architecturally you will be constrained to some degree. You just have to live with this because it's how the world works.

It's not a bad thing in some cases because some technologies are inherently better than others for particular use cases. So when you use that technology, and you use it the way the vendor intended you will get the most value out of it. If you try to avoid any and every vendor-specific nuance of the technology, you will not get the most value out of it. I get so frustrated when people say "We don't want to use X or Y technology because we want to be completely vendor agnostic." If you truly want to achieve that, then you will need to design your own silicon, servers, operating systems, programming languages, network protocols, applications, and even devices to view them on. I think you can see that this is not practical. So in the real world you have to accept that there is some level of vendor-specific interactions you are going to have to use.

I will be taking a very practical look at how to be as vendor agnostic as reasonably possible. You are after all working on creating *platform-independent* mobile apps and services. You don't want to be locked in to just the Android or Apple ecosystems. That will prove untenable in the long term. But at the core of this book is the idea that you can be very agile and deliver your apps on any platform now or in the future with minimal code changes and minimal disparate codebases. The server-side technologies don't have to be as vendor agnostic because they don't change much over time. Sure they change, and new vendors come up with new and interesting technologies and services for us to use, but it's insignificant when compared to the changes the mobile device landscape goes through.

So keep an open mind about that whole vendor lock-in thing. This goes for Open Source as well. Free and Open Source Software (FOSS) has gotten a lot of publicity for being for the people, of the people, and by the people. There are some fantastic open source libraries, databases and other bits out there like Node.js, Redis, and others. However, you need to use the right tool for the right job. What attracts most people to FOSS is the F part. It's Free! True, but you are relying on the good hearts of the FOSS contributors to support you if something breaks. That's perfectly fine if you are ok with "best effort" nature of some FOSS support.

Don't get me wrong, I'm not badmouthing FOSS, I'm just saying go into it with your eyes open.

- Understand the licensing implications of Creative Commons. In some cases if you use it, it means that your system has to be FOSS as well. So read which CC license the FOSS was released under.

- Make sure you know your support options and limitations. You may have to pay a third party company for support.

- Don't assume that the 10,000 eyes idea is valid because a lot of those eyes aren't trained software engineers and don't know what they are looking at. I've met countless FOSS advocates that don't know how to read code and have never contributed anything meaningful to FOSS. This is especially true for security issues. FOSS is not more secure because you can see the source code. All it means is that you have the source code. If you don't know what security vulnerabilities look like at the code level, then having the source code is just as useful as a screen door on a submarine from a security perspective.

- Look at the ongoing costs, not just the zero down payment. This is especially true of support, maintenance, updates, and stability. In many cases if you alter the source code and compile your own version, you are now in unsupported territory. Be mindful of this and plan your support options accordingly.

That being said, later in this book you will see me talking about and even recommending FOSS softwares like Hibernate, MongoDB, and others. Remember the golden rule.

You also want to be sure that you are choosing a vendor, cloud provider, technology, etc. that will be around for a while. There are many examples of technologies that people were banking on that a vendor suddenly decided they were going to turn off. In the early 2000, Google was famous for this. Does that mean you should avoid all things Google? No, of course not. But if something is in "beta" don't bet your production systems on it. Use technologies and vendors that have a track record of surviving the experimental phase. Does that mean you'll never get stung? No. Silverlight was a good example of that. It was a well-established and promoted technology by Microsoft. It was the way to replace Flash, but then, even after convincing all the mobile developers to build Silverlight apps for Windows Phone 7, Silverlight rode silently into the night. This kind of thing happens but you still need to pick a technology to build with. The more you go with proven stable technology the better.

You can and should still experiment with new technologies that come out. There will be some shiny new tech that shows up that will make your system amazing. Have a sand pit for this kind of thing and experiment. You won't always hit on a winner, but when you do, it will catapult your success to the next level. Just make sure that this new technology comes out of experimental/beta phase before you roll it into production.

With that understanding in mind, don't be afraid to pick a technology because a particular vendor produced it. One of our core tenants is be standards based. If that technology is as standards compliant as makes sense, then you should feel comfortable using it. If however, a particular technology is entirely proprietary, used by only that vendor, and does not have a standards-based interface you should think twice about it. For example, if a cloud vendor takes your virtual machine image, and converts it to their proprietary format, and does not offer a way for you to get your image back in it's original (or updated) condition, think twice about that.

If a technology stack has its own built-in proprietary messaging system that doesn't communicate via standard WS-*, RESTful, Message Queue, etc. protocols, you won't be able to decouple the layers and move things around. This will severely limit your agility and future proofing. Make sure that the technologies you chose are standards based or can communicate over standards-based mechanisms. Other than that, avoiding vendor lock-in is really just a matter of avoiding proprietary single-vendor technologies.

2.2 RECOMMENDED STANDARDS AND GUIDELINES

2.2.1 Respecting the Device

Most apps are on mobile devices now. These devices are designed for portability and as such run on batteries. They also have relatively small processing power and small permanent storage space. When designing your mobile app services, think about these limitations on your target devices and be kind to them.

Battery power is at a premium. Just ask any iPhone user who uses eagle-like vision to spot power points to run to before they unplug from the one they are on. Running your app drains this precious resource. So the more processing power your app requires, the more juice it takes. This can drain the device battery and earn your app a bad reputation amongst hipsters. There are a few key things that tend to drain power, location services, radio stack usage, camera, and CPU time.

You need to think about how you will use battery efficiently, and how to focus that use on your core operations, not fluff and things that don't benefit the user. For example, researchers at Purdue University [12] did some in-depth study of how popular apps use battery on mobile devices. In their study they found that Angry Birds, spent only 20% of its battery usage in actual game play and 45% of its energy usage was for tracking the user with the Flurry Analytics system. This means that the game spends more CPU time tracking users than it does actually playing the game. Does that matter?

This kind of thing matters when you go above the norm for app usage. Games are notoriously bad at battery conservation. This is usually because they keep the screen on and do a lot of CPU intensive work during play. If the users are having a good time with only 20% of your CPU time allocated to game play, and they don't notice the Flurry calls, you are probably fine. But if their phone battery is dead after a 10-minute game session, you've probably earned a trip to the delete pile.

The focus on how mobile app services can help you with battery conservation is by moving the processing power. For example, let's take a typical game. As mentioned above Flurry is a popular third party game and app stats engine. Your game makes calls to the Flurry API and those stats are uploaded to the Flurry servers. You can then analyze your users' performance, game play time data, and other things. This is where Angry Birds spends most of its time. But what if it could offload that for later processing?

Consider caching the information that would be sent to the user tracking system. Then between levels, or after character death, upload the batch of data to your mobile app service where the processing and uploading to third party services can be performed in the cloud, rather than from the device itself. This will save all of the CPU battery usage for your core purpose, and not third party libraries.

> *Rule of Thumb: Offload any CPU intensive workloads to the mobile app service whenever possible.*

There are some things you can't offload to a cloud service such as location information. The location of the cloud server will be relatively static and not the same as the user's location. Things that are very device contextual will need to be processed on the device, save your processing power for these kinds of things where possible.

If your app relies on location information from the device to provide the best experience for the user, try not to have it actively checking location constantly. Only refresh the location information when your app is active, and cache it until you need to know it again. Use battery intensive features like cameras, sensors, and network as little as possible while still providing a good user experience.

2.2.2 Respecting the Network

Another area that mobile apps tend to have limitations with is bandwidth. While some countries have Internet Service Providers that offer unlimited downloads and uploads, typically mobile carriers do not. In most countries around the world mobile data are at a premium. The more you use up that data plan, the less the user will like you. This comes back to one of our core distributed app tenants and another Rule of Thumb:

> *Rule of Thumb: Choose chunky network calls over chatty ones.*

This favors high-latency networks, as well as expensive data plans. Every time you establish a network connection, there are latency and bandwidth overheads associated with setting up the connection. So try to batch up data and send it in larger chunks rather than sending small bits of data frequently. This way, you avoid the overheads of creating and tearing down the network connections. With mobile apps you really don't want to open a network connection and leave it open if you can avoid it. This is a battery waster.

That being said if you apply the golden rule, you may have a multi-player game that needs to communicate in real time with other players, or some kind of system that is tracking live information from equipment, or stock levels. In these cases you may require a long-lived open web sockets connection. If that is what your system requires, then use it, but be cautious about leaving that connection open between games, or while the user is idle on a score board or store screen.

You want to keep the data sent and received as efficient as possible, especially in the case of the multi-player game example above. A case in point, don't send an entire dataset up and down the device, when only the deltas are required. Try to find network transmission data structures that convey a lot of information in as few bits as possible, byte arrays are a good choice and can be very versatile.

For example, if you were making a game and needed to know some basic and rapidly changing player data, you should use the most efficient data structure possible. Consider this simple set of data that you want to send to your mobile app service.

```
User:00013243
Location:x855.48,y122.33
Facing:270
Speed:10
```

This data come out to 67 bytes on disk in an ANSI encoded text file. That is 67 * 8 = 536 bits.

A JSON representation of this, which would be a very common choice for sending data to a RESTful web service, looks like this:

```
{"User":"00013243","Location":"x855.48,y122.33","Speed":"10",
    "Facing":"270"}
```

This is 76 bytes (608 bits) on disk and not surprisingly has a content length of 76 when sent via HTTP (Hypertext Transfer Protocol).

If you convert this to a byte array representation, where each piece of information is sent as part of the byte array you would do something like this.

You start with the data you want to transfer and its possible values. Let's say that you want to allow for 65,000 possible users which require 16 bits (I realize we'd probably have millions of users, but this makes the math easy), X and Y values from 0 to 1000 to two decimal places that represent our game map which requires 16 bit floating point numbers, speed from 0 to 10 which requires 4 bits and a compass facing from 0 to 359 which requires 9 bits. To represent these values, you'd have to allocate the following number of bits per piece of data.

User	16
X	16
Y	16
Speed	4
Facing	9
Total	61

All of the information can be transmitted in a single 64 bit integer value. This is a lot better. From 608 bits for transmitting the data in JSON format to 64 bits by transmitting the values in a byte array. You can still convey the same information and save 544 bits.

You could take this one step further and make smarter choices with your values. You can ensure our speeds only go to 8 instead of 10, and use integer values instead of floats for the X and Y coordinates. That could save another 13 bits, but at this stage, you're already much better off than with a JSON representation of the in-memory data structure. This is much more bandwidth friendly.

This might sound a bit oppressive or pedantic but let me put this in context. The goal of most mobile apps and games, and especially anything related to high-performance or high-volume input from IoT devices is to be able to handle really high volumes of incoming and outgoing data. Let's say you have an IoT app that receives data from ten million cars on the road. If you are receiving one message per second from each car that is ten million inbound messages per second. Every one bit you shave off that message, saves you 10 Mb/s in bandwidth.

Of course all of this need to follow the golden rule. If you are sending a lot of serialized objects, or sending textual data, byte arrays or bit maps might not be the ideal solution. JSON is better for that sort of thing. By the same context, ADO Datasets are great for conveying a lot of information about a set of relational data. However, it has a lot of overhead. So if you need to send a small change or retrieve simple data use REST/JSON. If the data are complex and you need to represent a relational dataset on the device, or if you need complex query capabilities directly from your service calls, Datasets may be the better choice. This is because of the metadata that describes the relationships and context for the data which allows you to do things like filters, selects, and "count" or "top" right from the query string. The general Rule of Thumb:

> *Rule of Thumb: Use the smallest data structure that makes sense, and allows for future growth.*

2.2.3 Communication Protocols

In this section, we'll talk about the standard data transmission formats. I'm not going to go into things like binary files, image formats, sound files, etc. I'm talking about the higher level protocols and formats. The lower level ones can be sent over the higher level ones such as HTTP and wrapped in XML, OData, and if you are creative in JSON. While this isn't so much about the standards themselves, but more about your decisions on which one to standardize on. It is likely that you will need to present two or perhaps three formats to the outside world, you should try to reduce this to one format on the server side for all of your internal communications and service calls.

For example, in the REST/WebAPI world you will probably use JSON as your default return type. You might then extend this to OData services, if you want to provide more flexible CRUD operations on your data. As an option for interaction with existing client apps or partner systems you may also have a WS-* based interface. There is a temptation when taking an approach like this to reduce functionality in the alternate data endpoints. If you primarily focus on REST/JSON, you might not put as much time and effort into an OData, or WS* interface and it might suffer developer neglect. Standardizing on one protocol where possible avoids this neglect for the most part.

The more data outputs you support, the more code there is to deal with and the more support people you need in place to help people with it. Large organizations such as Microsoft have been forced down this path because there are billions of computers running Windows in businesses that have some old piece of mission critical software that they depend on to run their business. There are lots of cases where there are encryption protocols that have been proven to be weak and no longer used, but someone out there, usually in large organizations, has some data encrypted with that protocol in a system that was designed decades ago that they simply can't live without. This has forced Microsoft into including those old outdated encryption protocols with current operating systems and the .NET language even though no one is creating new products with them. It is much to their credit that they didn't just abandon those organizations to their fate and did this, but this results in a real problem for support and more code being deployed that is necessary in most cases. You can't just upgrade the skills of your current support team and move them to support the new product, you have to keep people who are specialized in Windows XP and the 8-bit to 16-bit and 16-bit to 32-bit thunking layer [13] in position to handle support for these legacy systems.

With that in mind, you want to choose protocols and data types that are based on Industry standards, which are being taken up by the bulk of the industry. Choose ones that are extensible if possible such as text-based serialized formats like XML and JSON which most of the industry support. This is especially the case for externally facing endpoints. Ideally, you want your service layer to use the same formats when talking from service to service which gives you code reuse, and more team familiarity with the technology. This is not only for the dogfooding aspects, but also because a lot of other people outside your organization are using the same industry standard protocols.

Custom protocols are very tempting, especially in situations where you know exactly what will be sent and received every time. You can make the payload and protocol extremely efficient, but if you win the lottery and never come back to work, chances are your successor will have a hard time with your super-efficient and unreadable custom data transmission system. I have seen some organizations take this approach as a form of security where they thought that by using some obscure protocol and data standard it would make their systems harder to hack. While this is an interesting idea, it inconveniences you more than it does attackers. Using a plain text protocol over a channel such as SSL/TLS with mutual authentication [14] and high-level encryption is a much better approach.

There are some basic guidelines that help with this topic.

- Make your protocol and data format choices with a view to the future.
- Use ones that are easily converted from code-level objects to something easy to decode at the other end.
- Make sure that you can take advantage of compressed network protocols where possible, so you don't kill the users' bandwidth.
- Use protocols and formats that the industry is behind and avoid things that are pet projects, or overly proprietary if you can.

Next, we'll discuss some of the major transmission channels and data formats to get an idea, so you can make an informed choice.

2.2.3.1 TCP/IP   Transmission Control Protocol/Internet Protocol is one of the longest standing connection and communication channels in use by computers today. In fact, without it, there is no Internet, and in most cases, no networks. It has been used since before CAT6 cables and in fact since before Token Ring networks. (If you remember those, you are a pioneer). It is used almost exclusively as a lower level protocol for computer to computer communications.

TCP/IP is where IP addresses come from, and terms like IPV4 and IPV6. The IP part of TCP/IP defines the necessary protocols for using echo and network control systems such as ping. While this will work just fine for internal systems and LAN-based communications, as a direct protocol isn't not well suited for over the Internet communications by itself. This is mostly due to the limited number of addresses available on standard IPV4. With the prevalence of the Internet and mobile devices, the industry has run out of the 4.3 billion available IPV4 addresses. The industry has been feverishly working to fix this through new technologies such as Network Address Translation (NAT) and Classless Inter-Domain Routing (CIDR). More recently the successor to IPV4, IPV6 has been established and will support about 3.4×10^{38} addresses that is 3.4 with about 39 zeros behind it. If you're interested, Rob Elamb figured out how many that is, and how to say the number on his blog at http://elamb.org/howto-say-the-ipv6-number/. That may keep us going for a while until we start colonizing other planets and galaxies.

The issue is that not everyone supports IPV6 yet. This means it's not as straight forward as just dropping in an address. What if the address is assigned dynamically from a pool of them, you may not be able to connect to the same computer twice. This is similar to dynamic addressing in IPV4. The industry had to come up with a whole new system for dealing with dynamic IP addresses called domain name service which maps IP addresses too easy-to-understand names like www.mycompany.com. The hardest part was dealing with the explosion of Internet-connected devices.

The biggest issue is that the primary method of solving the IPV4 problem is NAT. This means that internally an organization will use an IP address range of whatever they want (usually something like 10.x.x.x), but so do other organizations. However, all of the Internet-facing endpoints are assigned an IP address from an issued range assigned just to them. Communications into that organization are mapped from the single externally facing IP address to the internally assigned 10.x.x.x address. This works very well as long as the internal networks are kept separate.

You have the problem that if you connect these two networks together through a VPN or a direct connection, you would end up with a lot of address conflicts. With TCP/IP, each address can only be assigned to one computer. So while TCP/IP direct connectivity works really well inside the network boundary of your organization, it's quite difficult to use once you leave the confines of your network.

As an example, it is very likely that if Alice's computer in Company A was to try and connect to the Bugle Boy's computer in Company B that it would try to reach an address of 10.100.100.1 since Company B assigned that address to his computer. However, this address is also assigned to Bob's computer in Company A and Alice will connect to the wrong computer. Alice's computer could potentially see two other computers both with an address of 10.100.100.1. This is called an IP address conflict and the system breaks down because the network routers don't know which computer with that address the message is destined for.

There are technologies in play such as NAT that make this possible. One of your companies' assigned external-facing TCP/IP address acts as the only IP address for your company that the outside world sees. So all traffic is routed to your public-facing IP address, and then the NAT router, using some extra information in the IP packets, translates the network traffic into the internal IP address that the communication is destined for. This works, and works pretty well, but with many devices, services, and cloud-based systems coming online, we're going to be pushing the limits of this as well because the externally facing IP addresses are primarily still IPV4 addresses.

Consider all of the services being deployed and made available from cloud providers. Amazon AWS, Microsoft Azure, Salesforce and others only have a finite number of Internet-facing IPV4 addresses they've been assigned to use. This problem is made worse by the fact that an IP address can be assigned to a computer or service, and if that service reboots or goes offline, it may be assigned a different IP address when it comes back if it wasn't assigned a static IP address. So using TCP/IP alone for your Internet-facing systems or cloud-based systems is not going to work.

You can use it for internal systems but you want to try to focus on using DNS-based addressing and machine/service names, not IP addresses. While IP addresses seem like a good approach internally because they are static and easy to code to, you can't use this effectively externally. You should create your systems so they will work with either external or internal endpoints. Many people have discovered this over the years as they have exposed more and more internal services to business partners and the Internet in the form of self-service mobile apps. So you may as well get used to designing apps this way from the ground up.

2.2.3.2 _HTTP_ The Internet as a whole began as little more than a way to transfer files, then eventually grew to include new groups, e-mail, and web pages. In the early days prior to 1990, FTP made up over one-third of all of the traffic on the Internet [15]. Within 5 years, World Wide Web traffic overtook FTP and hasn't looked back. This massive burst of information in the form of web pages created great swathes of new technologies and protocols. GIFs, JPEGs, HTML, URLs, ASP, JavaScript, RSS, ATOM, and most importantly, HTTP which all of these other technologies use to get passed around from web page to client and back again.

HTTP is a higher level application protocol for communicating between clients and servers or server to server. The IETF started work on it in 1999 and the more recent HTTP/1.1 in 2006. It went through almost 30 revisions over the years to its current format which is published in RFC 723X [16]. HTTP is defined in the RFC as a stateless application-level protocol for distributed, collaborative, hypertext information systems [17].

HTTP introduces some application level communication mechanisms and characteristics that operate above the level of network addresses and routing tables. It is designed to be stateless, operate on a Request/Response pattern using bi-directional transfer. It also allows for caching, intermediary (proxy) servers and the ability to negotiate the capabilities on each end of the connection.

This last part, capability negotiation, means that client apps can tell you what kind of data types they can accept and the fall back for them. This is done through Accept headers. For example, when I use my web browser to access the web page for RFC7230 these are the Accept headers my browser send the servers:

*Accept: text/html, application/xhtml+xml, */**
Accept-Language: en-US,en-AU;q=0.7,en;q=0.3
Accept-Encoding: gzip, deflate

It basically tells the server that my browser prefers to accept the text/html, but if you don't have that then application/xhtml+xml media types, but failing that pretty much anything */*. It also defines the languages I will accept which in this case is US English and Australian English. My browser can also deal with gzip and deflate compression schemes. This is obviously something that you wouldn't care about at the TCP/IP level, but is critical to application level communications. If you want more details on the inner workings of the HTTP 1.1 specification go ahead and check out RFC 723X. It's fascinating reading (your mileage may vary).

HTTP has become the ubiquitous application communication protocol of the web. If your systems are going to talk to anything outside your network boundary, you will need to use HTTP. REST, SOAP, WS-* all operate over HTTP. When considering bindings for your web services you should favor the HTTP-based bindings. In today's web technology-based world, this should be your default choice. Even though HTTP is a negotiation and text-based communications protocol that is not as efficient as pure TCP/IP, you need that to speak the language of mobile services and internet-connected devices.

2.2.3.3 REST While HTTP and HTML rule the web page world, when it comes to transferring data to mobile apps which may not be displayed in a browser, you use web services. Web services have traditionally been handled by various WS-*

and SOAP-based protocols. But as the industry has searched for efficient, and more importantly, easy to adopt systems RESTful web services have emerged as the leader in web service architectures. REST is not a protocol itself, it is really just a set of guidelines to create some best practices that allow you to design highly scalable web services. It really applies a purist philosophy to HTTP in some regards.

For example, in RESTful web services [18] the idea of statelessness is very well enforced. With normal HTTP-based web page operations the web world has worked very long and hard to overcome the idea of statelessness with state servers, cached data and a way to maintain state on the server-side to give the impression that each request/response interaction is actually just a single client app doing lots of long, multi-page interactions. With the REST approach, there is supposed to be zero client state maintained on the server-side. State is represented as a hyperlink in the responses sent to the client. These represent what the client can do from their current returned data, and where they have come from. This is officially referred to as HATEOAS, Hypermedia as the engine of application state.

One of the rules of REST that makes it very attractive to highly scalable web services is the Uniform Interface constraint [18]. This constraint essentially says that the endpoints and interactions for a REST-based service are all discoverable and defined by the URL the client calls to access the services. As we've been talking about, this provides the very beneficial ability to decouple the endpoint from the implementation. The URL endpoint is literally just a text string, and the implementation behind it can be anything. It can be moved, changed, swapped out for an external service or joined to another service without the client end ever knowing about it. The client side doesn't have to know anything about the physical implementation of the Service Layer or Data Structures other than what will be contained in the returned information.

The REST endpoint itself can describe the type of data transfer. For example let's say you were using WCF to define your web service, if you define your endpoints as

```
[ServiceContract]
public interface IHairDyeService
{
    [OperationContract]
    [WebGet(UriTemplate = "/")]
    ColorCollection GetAllColors();

    [OperationContract]
    [WebGet(UriTemplate = "/{color}")]
    ShadeCollection GetShadesForColor(string color);

    [OperationContract]
    [WebGet(UriTemplate = "/{color}/{shade}")]
    int GetStocklevelForShade(string color, string shade);
}
```

You can receive service calls as follows:

http://www.myservice.com/hairdyeservice/

This will return the list of colors: blond, brunette, red, black, blue

http://www.myservice.com/hairdyeservice/blue

This will return the list of shades for blue hair dye: Violet Blue, Green Blue, Ultramarine Blue Sapphire, Cornflower, Sky, Cobalt

Of course

http://www.myservice.com/hairdyeservice/blue/cornflower/

Returns 42.

On the back end if you map these URL indicators to a web service, or web api, or even some kind of text parser it doesn't matter. The mobile app itself can still call these URLs using these semantics and not have to worry about knowing what you've implemented on the back-end. They can however request data types such as JSON or XML. Ideally in the REST world, they do this by simply appending JSON or OData to the method call such as http://www.myservice.com/hairdyeservice/blue/JSON or http://www.myservice.com/hairdyeservice/blue/XML.

This ease of understanding and flexibility is also available for write operations using the PUT and POST HTTP verbs. This allows write operations to be executed against the same services.

```
[WebInvoke(UriTemplate = "/{color}/{shade}", Method = "POST")]
Shade AddShade(string color, string shade, Shade shade);
```

If you then call this with the POST verb rather than the default GET verb, you can supply the information to be posted to the method to add a new shade of blue hair dye. The POST request would look something like this:

```
POST /hairdyeservice/blue/shade HTTP/1.1
Host: myservice.com
name=PowderBlue&qty=150
```

For more information on building out these kinds of systems check the documentation for your favorite programming language or platform. It is important to note though that the REST URL endpoint will be the same regardless of what you have used to implement the service on the back end. This is the decoupling I mentioned. It is very important in future proofing your systems. There's a pretty good chance that standard URL conventions are going to be around for a very long time, but your implementation may change over time to accommodate new technologies and systems. Sticking to RESTful principles will help you out a lot in the long run.

Additional information:

REST: http://www.ics.uci.edu/~fielding/pubs/dissertation/rest_arch_style.htm

REST in .NET: https://msdn.microsoft.com/en-us/magazine/dd315413.aspx

REST in Java: http://docs.oracle.com/javaee/6/tutorial/doc/giepu.html

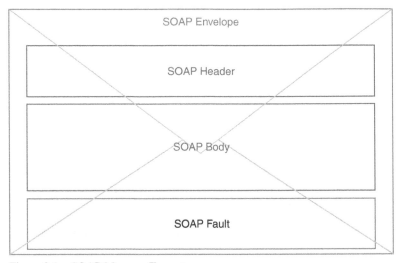

Figure 2.1   SOAP Message Format

2.2.3.4 SOAP   SOAP (Simple Object Access Protocol) has been used in web services for a considerable time. It was created in 1998 from a group at Microsoft [19] and is now maintained by the XML Protocol Working Group of the W3C (http://www.w3.org/2000/xp/Group/). It took the place of XML-RPC and is now in version 1.2 which was proposed as a W3C recommendation in April of 2007 [20]. SOAP messages can be sent over TCP, HTTP/S, and even SMTP if you want to e-mail them. SOAP is XML based and consists of a set of optional SOAP Header, a SOAP Body, and if it is a returned message SOAP Faults all wrapped up in a SOAP Envelope as depicted in Figure 2.1.

The header section in the SOAP XML, while optional, is of interest in a high-scale mobile app. One of the very useful features of SOAP is its ability to provide instructions to intermediate systems on its way to its final destination. These are referred to as forwarding intermediaries and active intermediaries. Forwarding intermediaries pretty much do what you'd expect. They check the headers for information on where or how to perform the forwarding. When they do this, they have to remove the headers that they have processed in addition to removing any headers that are not relayable while ensuring that they maintain all of the headers that are intended for the ultimate destination.

The SOAP specification also allows for digitally signing the headers, body or a combination of headers and the body. This can be useful for tagging and processing message paths. But when you combine this with active intermediaries, you can get some real versatility benefits. Your Service Interface Layer can be a forwarding intermediary or an active intermediary. It can not only accept and interpret the original SOAP request, but it can translate it, act on all or part of it, and break up the original message and send multiple downstream SOAP requests to different services as well.

You can extend SOAP through its extensibility model by creating SOAP Features for concerns such as reliability, security, routing, ingestion, and data translation among others. You can also define message exchange patterns such as one-way for the write side of a CQRS pattern, or read only for the read side. This is done through the SOAP Processing Model via SOAP headers which is used for processing single messages and the SOAP Protocol Binding Framework to describe the sending and receiving of messages by the nodes.

SOAP also uses SOAP Faults to return information to the caller when something goes wrong. This can be very valuable and is a bit more flexible than the standard HTTP style error codes returned by most RESTful web services. You can create custom errors to return to callers and put them in the subcode elements in the SOAP Fault messages you return. This can be used to provide additional information to the caller that can be customized to your services, and the method call they were invoking.

It's good to keep in mind that SOAP operates with the WS-* specifications. Therefore it can also provide message level security which can provide an extra layer of encryption beyond SSL/TLS. In most cases, since the messages are short lived you won't need this unless you are transmitting sensitive information that will be persisted. Additionally, SOAP via WS-Atomic Transactions can support two phase commits across services. I would argue though that unless you are doing things like banking transactions or something that requires synchronous dual confirmation for real time data consistency, you don't need this kind of overhead. If you require high speed and performance, this kind of approach will not suit your needs. You'll need to balance the decision between real time data consistency, and performance.

If you need this kind of extensibility and additional support for WS-* you can get that if you implement a SOAP-based system. Especially when it comes time to combine, move, or swap out services. SOAP is a well-supported protocol and has a lot of well-understood patterns of use available for it. That being said, SOAP is more used for large-scale and complex web services. I would recommend that a REST-based web services be your first port of call, but if you need the added flexibility, relay or active intermediaries, or WS-* support use SOAP-based web services.

Additional information:

SOAP Version 1.2 Primer: http://www.w3.org/TR/soap12-part0/

SOAP in .NET: https://msdn.microsoft.com/en-us/library/system.web.
    services.protocols.soaphttpclientprotocol(v=vs.110).aspx

SOAP in Java: http://docs.oracle.com/javaee/6/tutorial/doc/

SOAP in PHP: http://php.net/manual/en/book.soap.php

2.2.3.5 *OData*

OData (Open Data) is an OASIS standard protocol [21]. It is actually built on REST or more specifically designed to use REST interfaces for CRUD-like data operations. Whereas most RESTful web services tend to return just the answers to your service calls, OData can also provide information about the data model itself. It's closer to working with database access technologies just as ADO.NET or JDBC. The returned data is in JSON or Atom format although most often it is JSON. The OData protocol specifies that the service must use one or both of those formats at a minimum.

Often though when people think of RESTful web services, they think of OData as a separate thing. In actuality, OData is a set of guidelines to build RESTful

APIs that return JSON formatted data. But it extends the basic idea of RESTful web services to include queries like you'd use when connecting directly to a relational database.

Another aspect of OData that is particularly attractive to those who want to do everything from client proxies is the OData metadata. Each OData services creates a metadata document to describe the types, sets, functions, and actions that the OData service can understand. This tells the client how to interact with the OData service and what it can do with the entities in the service. This document is accompanied by a Service Document. The service document is the one that lists the functions, entity sets and singletons that can be retrieved from the service. The service document is used by clients to help navigate the Entity Data model (EDM) through HATEOAS by listing hypermedia links in the returned data to other information or functions that can be called from that point or state.

OData is also designed to not care about the back-end data store. So it suits our approach very well for any kind of services you want to create that include CRUD functionality, or where the client app may request information about the data model itself. OData uses an Entity Data Model which is used to describe the data you can retrieve from an OData service. You can build queries on the query string using parameters such as $search, $filter, $count, $orderby, $skip, and $top. OData even supports paging of data for very long lists of items.

Odata includes a comprehensive set of built-in filter operations such as *equal (eq), not equal (ne), greater than (gt), less than (lt), and, or, not* and *has* (for has flags). There are also arithmetic operators such as *add, sub, mul, div*, and *mod*. A complete listing can be found in Section 11.2.5.1.1: Built-In Filter Operations at http://docs. oasis-open.org/odata/odata/v4.0/odata-v4.0-part1-protocol.html. There is an extensive list of built-in query functions such as:

- String functions
  - contains
  - ends with
  - length
  - index of
  - to lower and to upper
- Date functions
  - year, day, month, hour, minute, second
  - date
  - time
  - now
  - mindatetime

There are also some Geo, Type, and Math functions available. A comprehensive list of these can be found in Section 11.2.5.1.2: Built-In Query Functions at http://docs.oasis-open.org/odata/odata/v4.0/odata-v4.0-part1-protocol.html.

Of course there is very robust create and update functionality built into the protocol as well. One of the more ideal features used to avoid data concurrency conflicts when performing updates is the built-in ETags. The ETag is a header that is returned with each entity. If you want to update that entity you must specify the ETag value in the If-match or If-None-Match header of the request to update or delete an entity. The Etag can be used as the data version to check and see if the data has been updated since the client retrieved the entity, or to ensure that that the entity the client is trying to update is the one that the request is actually for. This can be extended to enforce consistency across entities in the domain aggregate.

OData also supports asynchronous and batch requests. This is ideal for a write side CQRS implementation that accepts commands and processes them. The read side can be separated to handle the queries while the write side handles either synchronous or asynchronous insert, update and delete operations. You can also use the OData Change Sets to accomplish multiple updates or related changes in an atomic fashion.

If you were creating a service that was closer to a standard CRUD service, but wanted the simpler approach of REST, with the lighter payloads of JSON or Atom, and you were thinking of a CQRS/ES implementation, Odata is a pretty good choice. It is essentially a protocol built on modern web service architectural patterns and data formats. You'll have to tune your implementation, and enforce a good CQRS pattern if you want to maintain performance and high-volume mobile app services, but OData is up to the task. Just keep in mind that the OData JSON return format is much smaller than the AtomPub format. In some cases as much as 70% smaller [22]. So use the return format that is the most efficient.

Additional information:

OData Protocol: http://docs.oasis-open.org/odata/odata/v4.0/odata-v4.0-part1-protocol.html

OData in .NET: https://msdn.microsoft.com/en-us/library/cc668792(v=vs.110).aspx

OData in Java: http://olingo.apache.org/

OData in PHP: http://odataphp.codeplex.com/

2.2.4 Data Formats

In the previous section, we talked about the web services protocols that you can use to create your service layer. The two basic types we discussed both relied on HTTP to communicate with the broader Internet. But what about the data types you'll be returning to your service callers? What will you send to the device apps, or what will you send from service to service? If you are using SOAP, chances are you'll be using XML. If you use REST, you can chose between XML and JSON both of which are in common use in RESTful web services.

The ones you chose will depend on what you are trying to accomplish. In some cases, such as SOAP, the choice may be obvious, but for RESTful services, what you are trying to accomplish will determine which data format you use. You can return most formats from RESTful web services. Most of the time users expect JSON, or XML. Even binary return types can be contained in one of these formats. Obviously if you are returning information that will be displayed in a client, for binary types such as images and video you will likely return a URL to the resources for the client to display in the appropriate display control. This is especially the case in browser-based apps.

2.2.4.1 JSON   JSON is defined in ECMA-404, hopefully you can find it (that's a joke) [23]. JSON is very popular among web services, and mobile app services. It is a very light weight, only four pages, format for exchanging data between endpoints on a network or the Internet [23]. Its human readable and very easy for machines to read and write through various JSON serializers. It is also, very simple. It is essentially a collection of name:value pairs, or an ordered list of values. A value can be a string, number, true or false, null, an object or an array. You can nest these values as well which is useful for representing complex object hierarchies.

   JSON is the default format for returning data from OData calls, and is becoming the defacto standard for returning data from most RESTful services as well. Most programming languages have either built-in, or third party library support for serializing objects directly to and from JSON representation.

   We can use a serializer to convert these objects:

```
[DataContract]
[Serializable()]
public class HairDye
{
    [JsonProperty(PropertyName="Color") ]
    public string Color = "Blue";

    [JsonProperty(PropertyName = "Shades")]
    public Shade[] Shades = null;
}

[DataContract]
[Serializable()]
public class Shade
{
    [DataMember]
    public string Tint = "Blue";
    [DataMember]
    public int Quantity = 42;
}
```

Into this JSON representation (I added line feeds and tabs for readability)

```
{
  "Color":"Blue",
  "Shades":
  [
  {"Quantity":15,"Tint":"Cornflower"},
  {"Quantity":12,"Tint":"Sky"},
  {"Quantity":7,"Tint":"Midnight"},
  {"Quantity":19,"Tint":"Navy"},
  {"Quantity":42,"Tint":"Electric"}
  ]
}
```

This format is easily understood and transmitted by most services. It is easy to compress, and easy to deal with for AJAX style communications over web sockets or any other web-based communications. JSON's popularity is likely to continue into the mid to long term future due to its simplicity.

Protecting JSON data can be as easy as making sure your communications are over a VPN, SSL/TLS, or some kind of message level transport on your queuing system. You can also fully encrypt JSON documents for storage.

There are several NoSQL databases now that are designed to work directly with JSON serialized data. These are often referred to as Document Databases. Normally these use some kind of key and then store the serialized JSON document with the key. There are many such document databases. Some are stand-alone like RavenDB (http://ravendb.net/), MongoDB (http://www.mongodb.org/), as well as cloud-based services such as Amazon DynamoDB (http://aws.amazon.com/dynamodb/) and Microsoft Azure DocumentDB (http://azure.microsoft.com/en-us/services/documentdb/).

These document databases allow indexing, searching of contents and metadata, as well as complex query operations on the contents of the JSON documents. For example you can use standard SQL style queries to find items:

```
SELECT *
FROM HairDyes hd
WHERE hd.color = "Blue"
```

This query was set up on Microsoft Azure DocumentDB and returns all of the hair dyes with a color value of Blue. A full request for this same query on the REST API would look like this:

```
POST https://rhhairdyes.documents.azure.com/dbs/Z5tmAA==/
    colls/Z5tmAOfUegA=/docs/ HTTP/1.1
x-ms-documentdb-isquery: True
x-ms-date: Mon, 18 Aug 2014 13:05:49 GMT
```

```
authorization: type%3dmaster%26ver%3d1.0%26sig%3dkOU%
    2bBn2vkvIlHypfE8AA3uuf8zKjLwdrx7hwn0YGQ%3d
x-ms-version: 2014-08-21
Accept: application/json
Content-Type: application/query+json
Host: rhhairdyes.documents.azure.com
Content-Length: 50


{
    query: "SELECT * FROM HairDyes WHERE (root.hairdye.color
    = 'Blue')",
    parameters: []
}
```

All of this is based on storing JSON documents. If this is your primary data transfer format, then using a document database for storage, searching and retrieval may be very easy to implement. Just keep in mind that while you can get some NoSQL databases that do JSON document style operations, most of them don't support industrial strength relational database features such as analytics, encryption, automated mirroring, etc. Some do, and some don't be sure to read the package contents and follow all instructions.

As you can see JSON is fully supported from the top to the bottom of the application stack. Readable by humans and machines, easy to compress and transport, easy to get from storage into code level objects and back. It is a very versatile format and most importantly for high-volume mobile app services it's very lightweight.

2.2.4.2 XML Most of the app development world knows what XML is. It is part of the Standards General Markup Language (SGML; ISO 8879:1986) and was released to the public in November 1996 by the W3C working group [24]. It is used in everything from .NET configuration files, to data persistence though XML serialization. There are even some workflow integration systems such as Microsoft BizTalk that are designed to work with XML documents and translate them from one endpoint format to another with XSLT. XML is as pervasive on the web as HTML is. Apple iWork file formats and Modern Microsoft Office Open XML file formats are completely XML based. XMPP even uses it as its primary communication protocol. Common news feeds RSS and Atom are also XML based.

It is also highly customizable and extensible. With custom XML namespaces you can create your own definition of the types of data nodes in the XML document, how they relate to each other and what the values represent. XML is an extensible as JSON is simple. There is a plethora of information on XML, how to use it and how it works so we won't' go into detail here. But we will discuss some of the aspects you may want to consider for your mobile app services.

Firstly, XML was designed in the time of 14,400 baud modems. It was as efficient as it could be but over time as the extensions and commonality of it grew it became more than just a data transfer format. As the world fell in love with mobile devices and the bandwidth restrictions kicked in, XML fell out of favor for more lightweight formats such as JSON. For example, if we serialize our HairDye class from earlier in XML we get the following:

```
{<?xml version="1.0" encoding="utf-16"?>
<HairDye xmlns:xsi="http://www.w3.org/2001/XMLSchema-
instance" xmlns:xsd="http://www.w3.org/2001/XMLSchema">
  <Color>Blue</Color>
  <Shades>
    <Shade>
      <Tint>Cornflower</Tint>
      <Quantity>15</Quantity>
    </Shade>
    <Shade>
      <Tint>Sky</Tint>
      <Quantity>12</Quantity>
    </Shade>
    <Shade>
      <Tint>Midnight</Tint>
      <Quantity>7</Quantity>
    </Shade>
    <Shade>
      <Tint>Navy</Tint>
      <Quantity>19</Quantity>
    </Shade>
    <Shade>
      <Tint>Electric</Tint>
      <Quantity>42</Quantity>
    </Shade>
  </Shades>
</HairDye>}
```

As you can see, while it's very descriptive, it is also considerably larger than the JSON version. It is in fact almost three times the size of the JSON representation. While this is fine running on a LAN, or very fast home Internet connection, for mobile devices this kind of overhead tends to add up.

Is it worth the overhead? The answer, unfortunately, is it depends. With this overhead comes a lot of flexibility. Each XML node can have different attributes that can be used to provide contextual information. Simple examples are declaring node types. If we just decorated the Shades class like this: (other attributes left in for completeness)

```
[DataContract]
[Serializable()]
[XmlType("Tint")]
public class Shade
{
    [DataMember]
    public string Tint = "";
    [DataMember]
    public int Quantity = 0;
}
```

Then we get this result:

```
{<?xml version="1.0" encoding="utf-16"?>
<HairDye xmlns:xsi="http://www.w3.org/2001/XMLSchema-
instance" xmlns:xsd="http://www.w3.org/2001/XMLSchema">
  <Color>Blue</Color>
  <Shades>
    <Tint>
      <Tint>Cornflower</Tint>
      <Quantity>15</Quantity>
    </Tint>
    <Tint>
      <Tint>Sky</Tint>
      <Quantity>12</Quantity>
    </Tint>
    <Tint>
      <Tint>Midnight</Tint>
      <Quantity>7</Quantity>
    </Tint>
    <Tint>
      <Tint>Navy</Tint>
      <Quantity>19</Quantity>
    </Tint>
    <Tint>
      <Tint>Electric</Tint>
      <Quantity>42</Quantity>
    </Tint>
  </Shades>
</HairDye>}
```

We can even convert properties into XML attributes like this:

```
[DataContract]
[Serializable()]
[XmlType("Tint")]
public class Shade
{
    [DataMember]
   [XmlAttribute]
    public string Tint = "";
    [DataMember]
    public int Quantity = 0;
}
```

Then when we serialize it, we get the following:

```
{<?xml version="1.0" encoding="utf-16"?>
<HairDye xmlns:xsi="http://www.w3.org/2001/XMLSchema-
instance" xmlns:xsd="http://www.w3.org/2001/XMLSchema">
  <Color>Blue</Color>
  <Shades>
    <Tint Tint="Cornflower">
      <Quantity>15</Quantity>
    </Tint>
    <Tint Tint="Sky">
      <Quantity>12</Quantity>
    </Tint>
    <Tint Tint="Midnight">
      <Quantity>7</Quantity>
    </Tint>
    <Tint Tint="Navy">
      <Quantity>19</Quantity>
    </Tint>
    <Tint Tint="Electric">
      <Quantity>42</Quantity>
    </Tint>
  </Shades>
</HairDye>}
```

It's also possible to declare pretty much all of the properties of the class as attributes if you want to although that's not exactly best practice. For example I wouldn't recommend serializing an array as an attribute. One ancillary benefit of this is that is does reduce the output size somewhat. I would say though that under normal circumstances it won't reduce it enough to get to JSON efficiency.

What you do have though is versatility and power. These kinds of constructs allow you to be very explicit in your data types. You can include subclasses, have very fine grained control of serialization with advanced attributes, and functions as well as take advantage of built-in serializers in almost every programming language. XML is also the default SOAP data format. Perhaps the biggest advantage XML has over JSON is how easy it is to serialize binary data and store it in a CDATA section in the serialized XML. At the time of this writing, this was somewhat difficult to do in JSON.

You need to consider your requirements when deciding on the data format. If you want lightweight speed and simplicity you will probably go with JSON. If you need more complex fine grained control, or need to express information in attributes as well as nested values, or you need to pack binary data into the serialized objects, you will probably go with XML. If you are using SOAP-based services, or you want to define your own vocabulary, XML is the well supported for those scenarios. If you are using more REST-based service, the default tends to be JSON.

I would recommend that when you are doing real time AJAX (despite the X in AJAX standing for XML) calls or web sockets where speed and responsiveness I would use JSON. JavaScript understands JSON natively. While there have been calls to deprecate XML in favor of JSON [25] I think this is an overreaction. You need to use the right tool for the right job. Remember the golden rule, you need to do what makes the most sense for your situation. Don't let architectural purism get in the way of reality and practical application.

2.2.4.3 Binary Binary serialization is quite efficient from a size standpoint. That being said when serialized as files on disk there is extra header information, that is put into the file. When I wrote the various serialized outputs of our BlueHairDye from the previous examples to disk I got the results shown in Table 2.1.

It's rather curious that when compressed (zipped in this case) the compression information actually made the JSON file lager, but the other two, particularly the XML file compressed pretty well. However, even compressed neither the binary nor the XML file was smaller than the uncompressed JSON. While this seems to make JSON a clear winner, the larger the files the less the header information, as a percentage of overall file size, adds to the size of a binary file. The header information pretty much remains the same regardless of the size of the file, so you will reach a point where a compressed binary will be smaller than a similarly compressed JSON file. However, with XML the structure is such that you don't get that advantage as

TABLE 2.1   Serialized File Size Comparison

File	Bytes	Zipped
hairDye.json	189	239
hairDye.bin	446	378
hairDye.xml	563	356

the file grows. XML does compress rather well though due to all of the duplicated symbols. You can get very good XML compression out of things like XMILL which is probably the most efficient XML compression systems in use [26].

You can also use other data formats if you need to such as Smile (http://wiki.fasterxml.com/SmileFormatSpec), which is a JSON compatible binary format, however the language support for Smile is lacking compared to the more standard binary, JSON, and XML. This makes Smile a poor cross-platform choice. Internally, it may work fine, and if you are the only one writing the client/app side, you can probably get away with it if you write your own parser in the language of your choice. Similarly there is BSON (http://bsonspec.org/) and BJSON (http://bjson.org/) which are binary JSON formats. BSON does have some momentum and there are several implementations of it. It is used in MongoDB and there are processors for Java, C#, and Objective-C that you can use.

There have even been suggestions of using a transposition method to create arrays of similar data rather than repeat the field names in the JSON data [27]. While this will yield slightly smaller sizes, you only really get efficiency out of it for very large JSON data. The complexity of processing the transposition on each end outweighs the efficiency of the smaller sizes.

Another thing you'll need to contend with when sending binary data to other systems is that you will likely have to enclose it in an XML CDATA section or some other wrapper. While it is lower level and tightly packed, it is harder to work with using web technologies. It is also not human readable, and it is not easy to discover its uses when provided over a web service without guidance. If you control both ends of the communications channel, then binary might work, but it is very difficult for third parties or partners to integrate with. Like most serialization formats, binary serialization is designed for storage on disk and more so for socket-level network streams.

The issue with binary centers around the proprietary nature of it. The Web is more and more about discovery, mashups, big data, and services you can take advantage of. If you serialize data in binary, you are the only one that really understands it without some effort involved. If this is your intent, then great, but in a world of mobile app services, this isn't the approach you want to take if you want interoperability, extensibility, or any kind of future proofing.

If you already have a lot of legacy binary serialized files, or communication channels in house that you are trying to get exposed as services, use the SIL to translate your binary into XML or JSON. This allows you to utilize existing systems and internal services without re-writing from the ground up. We've added the SIL and DAL into the traditional architecture stack to provide this flexibility. It also allows you to continue to use in house formats in house while being able to use Internet standard formats externally.

Something else of note here, several people I've encountered have considered binary serialization, or in fact using native languages like C and C++ to be a security measure because the code that is deployed is binary in nature. That is to say not very human-readable. They have also extended this view to their serialized data, and messages sent across the wire. Binary compiled DLLs, and binary serialized data is not protected from attackers. Binary serialization is not encryption. Yes, it makes it so that your generic script kiddie can't just crack open the DLL with something like Reflector (http://www.red-gate.com/products/dotnet-development/reflector/) and read your top secret algorithms or data. However, if that is the level of attacker you are worried about, use something like Dotfuscator (http://www.preemptive.com/products/dotfuscator/overview) or equivalent for your language of choice to obscure it. More importantly put all of the code you want to protect in a service in the cloud behind an authenticating and validating SIL. If you are worried about actual attackers, they'll use tools like IDA Pro (https://www.hex-rays.com/index.shtml) to take apart your binaries and read the code like it was plain text. Bottom line is, "security" should not weigh heavily into your decision to use binary as a data format.

To protect your serialized JSON or XMl data in transit, use standard SSL/TLS, IPSec tunnels, and message level encryption. Don't rely on the security by obscurity which is what you are doing if you think binary serialization is a protection mechanism.

2.2.5 Mobile User Experience Guidelines

There are a lot of resources on the internet, classes you can take, and books you can read about the user experience (UX) guidelines for mobile device apps. There are a few things to consider overall though each platform will have slightly different UX guidelines. Some devices want menus on the top, some want them on the bottom. In either case, you will need to look at the specific platform's UX guidelines to be sure you are designing a high-quality user interface.

That being said, since this book is concerned with user experience from a services angle, we will discuss how to make the app as responsive as possible, and provide the user the best experience while taking advantage of high-powered services on the back-end. This is primarily focused on the responsive and efficient services that allow the apps to be kind to the device while providing the user a snappy responsive interface that seemingly knows what the user wants to do next.

For mobile users the satisfaction of the user experience is inversely proportional to the response time of the app. Your aps should be snappy, and easy to figure out. To provide a good user experience there are a few things that a lot of the industry has settled on [28] that make a good user experience. There is a balancing act illustrated in Figure 2.2.

2.2.5.1 *Start Up Fast*   Naturally you want the user to think your app is always loaded and ready to go. So if you spend a lot of time on startup making several service calls, or downloading a huge amount of data to display while the user starts as an indefinite wait cursor, they are going to think the app has crashed or the mobile OS will kill your app for not starting up in time. This is where you need to balance the use of locally cached data, with an acceptable timeout period before the app has to fetch fresh data.

What you should do is get the app up and running with the last known state. This can either be a static screen or one that uses data which has been saved to local storage. Then check for new data in the background as the user is looking over the last bit of interactions they had with your app. If there isn't any new data move on,

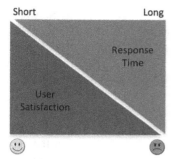

Figure 2.2   User Satisfaction versus Response Time

if there is you can fetch it if required. This initial service call should be as light as possible. You can make more calls later that lazy load more data as needed.

Ensure your service calls can identify the most current set of data so that it can shortcut the data request process. You can even use various HTTP conditional requests, or ETag version information. Using the HTTP Conditional Requests you would only retrieve data if it was modified after the specified time. This requires two steps, first the server sends the client a Last-Modified value such as:

```
Last-Modified: Tue, 03 Mar 2015 18:42:57 GMT
```

Then each time the client makes a request to the server after that, it includes the If-Modified-Since request header:

```
If-Modified-Since: Tue, 03 Mar 2015 18:42:57 GMT
```

If the server validates that the resource does not have a more recent modified date, then it hasn't changed since the copy the client has was retrieved. So instead the server sends an HTTP Response code of 304 Not Modified and the client doesn't have to retrieve the data.

The same type of operation works with ETags and the If-None-Match request header. On the first response to the client, the server sends the ETag value.

```
ETag: "FD139BBA27FDADADCCAE446E65BBE798"
```

Then the client can sent just the ETag back in the If-None-Match header to see if this still exists.

```
If-None-Match: "FD139BBA27FDADADCCAE446E65BBE798"
```

Since ETags are hashes of the actual data or object you are checking. So when you ask for the If-None-Match is evaluated, what you are saying is, if the ETag (hash) I have, no longer matches anything you have, then the data I have must have been changed, therefore I should ask for the new data.

Using these techniques you can very quickly tell if the data on the client is stale and needs to be refreshed. Ideally, if it is suitable to your application, you can wait until there is new data then ask the user if they want to refresh their data. This way you only incur the network and download hit when the user requests it, and they are expecting it and know why the app is spinning. This will allow you to get the app up and running and show the user something useful before taking a while on service calls that may be downloading larger sets of data.

2.2.5.2 Use REST HATEOAS The idea behind HATEOAS is to use hypermedia to essentially show the users what actions are available from the point they are now. HATEOAS is dynamic, and you can add the relative links into any response as the services and capabilities of your service grow or change. For example if a user has been using your app and service to purchase blue hair dye, they have probably been given options when viewing a particular shade of buying it or going to the next shade. The links to purchase, and move to the next or previous shades are supplied in relative links in the JSON response from your RESTful web service. Let's say that up to now you've had relative links in your responses that look like this:

```
<link rel="next" href="/dye/blue/shade/cornflower" />
<link rel="prevoius" href="/dye/blue/shade/navy " />
<link rel="buy" href="/cart/blue/electric" />
```

Now you want to break into social media and get a really good customer interaction path going. So you add a Share on social media feature and a Provide Feedback capability. Links to the shade they are looking at, and sharing this on Facebook or Twitter can be included in the RESTful response as relative links. The same goes for the Feedback link.

```
<link rel="next" href="/dye/blue/shade/cornflower" />
<link rel="prevoius" href="/dye/blue/shade/navy " />
<link rel="buy" href="/cart/blue/electric" />
<link rel="Facebook" href="/facebook/blue/electric" />
<link rel="Twitter" href="/twitter/blue/electric" />
<link rel="Feedback" href="/feedback/blue/electric" />
```

If your app just has navigation section on it, and it goes through all of the relative links building buttons for the possible actions, when you start including these two new features, they magically appear for the users as soon as they refresh the screen.

It simplifies the interaction model, and allows the insertion or removal of features very easily. If users can see new features and options popping up in the app without having to update or download a new version, that is a very cool thing.

2.2.5.3 *Integrate with Their World*
I admit this one is not entirely about the services and something of a pet peeve of mine but there is a strong service connection. Use your services to integrate with what is important to your users—social media, shopping trends, news, and entertainment. Provide your users ways to integrate their app experience with their connected world.

Let me give you an analogy to explain what I mean. In my opinion of all the great things Apple has done with the iPad this is something that really lets it down. Ironically enough Microsoft got the idea very right when they started with Windows 8/8.1 and 10. The device that people carry around with them is a lifestyle choice. Some people carry iPads, some carry Galaxy Notes some carry Surface Pro 3s. In each and every case, they chose a device that was suited to how they operate and what they want the device for, or what they could afford. Your app needs to integrate with that lifestyle choice.

With iPads there was no consistent way to interact with apps. Some looked like they were designed by Apple, some looked like they were designed by a blind potato farmer. Sometimes they had menus on the bottom, sometimes on top. Some used the "hamburger" button, some didn't. Each app may have had its own internal settings, or you may have been able to set them through the iPad settings. So when you wanted to change an apps settings, you had to hunt around until you found it. It was even different from app to app of the same kind. For example one news reader let you set the feeds from a tab at the top, another news reader used a setting under the "hamburger" button and yet another set them from the iPad settings. So even this simple thing because cumbersome and a game of hide and seek from app to app.

This makes the iPad feel like a briefcase. In a briefcase you might have a calculator, a ruler, the newspaper, your phone, etc. But they are all just dumped into the briefcase. Each item in the briefcase is its own individual thing with little interaction with the briefcase or other things in the briefcase. The iPad was just a bag to carry apps around in. You might have the sexiest briefcase in the world, but it would be full of things that looked like your grandmother's junk drawer.

With Windows 8.x+ devices Microsoft laid out a set of interaction guidelines that while strict, forced people into the minimalistic modern app style. App interaction styles were similar. They were predictable not only from app to app of the same type, but between different apps entirely. Let's take Settings for example. It didn't matter if you were changing the settings on the device, or an app, you swiped in from the right to open the charms bar, tapped Settings and changed what you wanted to change. It didn't matter who wrote the app, or what kind of app it was, the experience was consistent across everything. That predictability was comforting (after you discovered it and got used to it).

This simple integration into the OS made everything work together. The Windows 8.x+ devices were not seen as a briefcase. When you installed an app, it added features and functionality to your lifestyle device. A news reader provided an integrated way to read news feeds, the calculator allowed you to use your device to figure out how much money you wasted stocking up on Red hair dye. But they all did it in a consistent look, feel and interaction style. Every app felt like an extension of the device.

Integrate with their world. Be contextual with the user interactions and data your services present. Use your services to connect them to things they love most. If you can do that, you'll have chart topping apps.

2.2.5.4 *Cross-Device Data*

Part of integrating into the user's world is making sure that you present a consistent experience across devices that they may be using your apps on. While the apps themselves may be different on each platform, or form factor, what the app does and the data it works with should be the same and available across all of them. The W3C calls this the Delivery Context [28] of the app. You need to be aware of the Delivery Context to ensure you are being data wise.

The way a user interacts with your app on a desktop computer or laptop where screen real-estate is plentiful is as important as an efficient scaled down UI on phablet or phone screens. In either case, you need to be able to provide the data from your services that is most appropriate to the interaction style of the device. You don't want huge complex spreadsheets showing up on a phone, and by the same token you don't want a huge screen with filler background that has a small 640×1136 area of user interaction space floating in the middle of the screen.

That being said, you do want the data they present to be the same yet contextualized to the form factor. This requires your service to be able to present data to all device types. To do this, ensure that all of your client devices are calling the same service. Of course you say, that is what we've been talking about all this time. But more importantly, make sure your services are returning the amount of information appropriate to the device so the device isn't trying to display a three hundred row dataset on a phone. When the user at the office enters information from their PC in a large screen format client app, that information should be displayed as soon as possible on the phone in a format that makes sense for the smaller screen.

This is where making sure your app is synchronizing the data, reporting data synchronization updates, and refreshing data whenever feasible. In a CQRS/ES system, as long as the devices are drawing from the read side, the data presented to them will be eventually consistent. In more direct CRUD operations, they may be strongly (immediately) consistent. There is more on this in 6.5 where we discuss the CQRS pattern.

You will also have to consider the user interacting with the app and its associated services through multiple different devices at the same time. While this will not happen a lot, it probably happens more than you think. So ensure that your consistency rules and cached data refreshes are well ironed out. This is because in this situation you will most certainly get consistency and replication conflicts.

The trick is to make sure that you are not only presenting consistent, synchronized data, but that you are only sending the data to the client that it needs. The trick is another Rule of Thumb:

Rule of Thumb: Send as much data as required, but no more.

If your service is being called by your desktop app that is capable of displaying all of the data viewable by the user, you will send all the data they would have to do pretty much everything the system does. But if you are sending in formation to a phone, it's only going to display the top level information or most critical information, and the interactions may be restricted to what you can do with a thumb and an LTE connection. Restricting the information sent has several other benefits.

- You are automatically being bandwidth friendly.
- You are ensuring that there is only as much data on the device that needs to be which is better for security.
- The less data you send, the lower the chance of conflicts.
- The less data you send the less you have to cache in local storage making you device friendly.

You can accomplish this with simple keys sent from the device that describe its Delivery Context. Ideally you want to do this detection at the server end so you can tailor the service response. This is especially the case for web-based clients although for web clients you can use JavaScript and CSS Media Types/Queries on the client to get information and send it to the server.

At the HTTP level you can use the Accept and User-Agent headers. Keep in mind that User-Agent can be set to anything the user wants from a web browser inter-action, but a mobile app is more predictable and you can in fact set this to whatever you want from your first party apps. By recording this information keyed with a par-ticular user, you can build a device description repository [29] which contains the list of device/user combinations and the associated relevant device characteristics. This will give you pretty a pretty good understanding of the Delivery Context you need to operate in. Once you know what that context is, operate in it efficiently and contextually.

2.2.6 Authentication

2.2.6.1 *Claims-Based Identity*   Before we get into the authentication standards used in modern apps, I want to clarify something to make the discussion easier. Authentication and Authorization are not the same thing.

Authentication is proving a user is who they claim to be. Authorization is deter-mining that they are allowed to do whatever they are trying to do. You cannot autho-rize a user until they have been authenticated. By the same token (you see what I did there) you can have an authenticated user that is not authorized to do anything.

Authentication is a concern across all aspects of computing. Amongst the many concerns when it comes to authenticating users is positive user identification, and avoiding forcing the user to create yet another identity. Another aspect of this that developers often overlook is that if you create your own user store that you use to authenticate a user, such as storing a unique username and password, there are a lot of security concerns that go along with that. You are now holding information that could be considered a high-value target for attackers. Hopefully your app and services are popular enough that everyone uses them. Attackers may want to get at your user information store to find out what a particular user uses for their password, because typical users use the same password across everything.

This means if an attacker has trouble cracking a user's Facebook, Microsoft, iCloud, or Gmail password, they might be tempted to attack your system to find out what it is. Consequently you have to implement the same high levels of protection that Facebook, Microsoft, Apple, and Google do. Are you up to the task? This idea of federated identity and claims-based authentication tries to solve that problem for you by letting someone else hold the user authentication information and perform the actual authentication process. This is referred to as claims-based identity.

The industry seems to be converging on the idea of claims-based identity to solve this issue. This involves what is referred to as Federated Identity. Federated Identity is where one provider, say Google, Facebook or Microsoft, can authenticate someone, and issue a digitally signed token. That token proves that this trusted third party validated that the users is who they say they are. All you have to do is accept and trust that token then make your authorization decisions based on what you want to allow that user to do.

Here's an example most of us can relate to in the real world. Buying a drink at a bar. This involves three things. The Bar who is the Relying Party, the Driver's License issuer who is the Identity Provider, and you the dude/dudette that wants a cold one. When you go into a bar, the bartender asks you for your ID, this is usually your driver's license. Your driver's license acts as a security token issued by the department of motor vehicles. It has claims on it. Your name, address, date of birth and the class of vehicle you are allowed to drive. It also has a "signature" of the issuer in the form of holograms, and anti-fraud mechanisms, plus an expiration date. The bartender doesn't have to know who you are or keep a list of all the people that enter the bar (assuming this is not a mafia bar). He trusts the driver's license to tell him you are old enough to buy a drink. He trusts that the department of motor vehicles has already validated that you are who you say you are and that you were born on the date on your license. All he has to do is trust the DMV did its job and trust the license.

In the app context, your app is the relying party (the bartender), your users will be issued tokens (the driver's license) from the identity providers of your choice like Google, Microsoft, Apple, Facebook, (the department of motor vehicles). That token will have claims in it such as who issued it, who the principal (user) is that it was issued for, and expiry time, and a signature. You just have to validate the signature then parse the token to find the claim you want. Once you do that, you can trust that the user has been authenticated by the trusted token issuer. You can now make your authorization decisions based on that authenticated user.

2.2.6.2 The Standards Two of the main claims-based identity systems in this area are the WS-Security (https://www.oasis-open.org/committees/tc_home.php?wg_abbrev=wss) / SAML (https://wiki.oasis-open.org/security/FrontPage) system and OAuth2 (http://oauth.net/) The WS-Security protocol with the SAML (Security Assertion Markup Language) payload has been around for a while and was in fact last updated in November of 2006. This has been part of the WS-* standards for some time and most systems can accommodate it. Many enterprise directories such as Oracle Identity Manager and Microsoft Active Directory inherently understand the WS-* protocol and SAML packages. It is the primary authentication mechanism for most SOAP-based web service calls in common use. However, this has proven to be a very heavy system. SAML assertions tend to be very large XML structures in comparison to the OAuth2 tokens. This makes them pretty hard on mobile app systems.

The OAuth 2.0 tokens are typically only about 5 lines and can fit into an HTTP authentication header. This makes them very attractive for web services, and mobile devices. OAuth 2.0 is also gaining large support across the industry. At the time of this writing Microsoft, Facebook, Twitter, Google, Yahoo! and others have implemented support for the OAuth 2.0 protocol.

OpenID (http://openid.net/) is an identity extension to the OAuth 2.0 protocol. In addition to allowing client apps and services to verify the identity of a user, it also allows the client to get basic profile information about the user. OpenID is in use with things like Google, Facebook, and other systems that let you get profile information from users that are registered with their respective systems. By using OpenID, you can not only have the larger organizations like Microsoft, Google and Facebook handle the authentication, but you can get additional information on those users from their respective profiles.

If you hadn't thought about using claims-based identity, you really should be. The more security issues you can offload to organizations that are better able to handle it, the better. By combining bandwidth friendly tokens, with widely adopted protocols and standards, your services will have a long and well-supported future.

2.2.7 Dealing with Offline and Partially Connected Devices

When you are designing your app services, you need to consider if you are going to design for always connected or partially connected scenarios. There is a considerable difference which means you need to make some data structure and architectural decisions upfront. Allowing the client app to operate when offline is a big advantage. Especially when the app will be used in locations where connectivity is non-existent such as surveying, farming, mines, or on the ocean. This is a pretty significant design decision. Let's consider the scenarios.

Online Required: These are apps that need to be online to function. Video streaming apps, web browsers, live IP telephony apps, etc. If they do not have a connection to the Internet, they can't do what they are made to do.

Online Preferred: These are apps such as asynchronous messaging apps, e-mail, news readers and other apps that provide their best experience when online but are quite functional offline. You can read cached news stories, read and write e-mail, and send messages to be delivered later. The waiting outgoing transactions will be sent and cached data refreshed when online.

Online Regularly: These are apps that expect to be online on a regular basis to sync or check in but are designed to operate offline after the sync. These are usually field work force type apps such as work order apps, data collection apps, or anything that expects to work offline and then sync updates in a batch when online.

Online Optional: These are apps that after the initial installation can operate fully without ever connecting again, but may require online connections for updates, or an enhanced experience. Things like mapping apps that download maps for use offline, or single player games that don't communicate during the game but can download updates and new content at the player's request.

Online Not Required: These are more traditional apps that were intended to work stand-alone. CAD or drawing apps, spreadsheets, word processors, and similar apps that don't need contact with the outside world. Since we are discussing app services and dealing with partially connected scenarios, we'll limit this part of the discussion to Online Preferred, Online Regularly, and Online Optional apps.

2.2.7.1 Partially Connected Approach

There are typically two approaches to dealing with partially connected apps. You can store a local copy of the dataset on the client device, or you can just batch up the requests and call the service when connected.

If you store a local copy of the dataset, the client can perform all of the normal CRUD operations it would expect to, and the database itself handles sorting out the conflicts through merging or a last in wins approach whichever is suitable. This would be commonly found in field workers doing data collection, or processing work orders, etc. This works well because it's less code you have to write to deal with conflict resolution, and holding the queued messages. However, it does mean that you have to have a local database on the client. There are many such databases suitable for this kind of work now that are small enough to operate on the limited resources of a mobile phone or tablet. You can use things like Web SQL, which is in use but no longer actively maintained, SQL Lite, SQL Server Express, and eventually Indexed DB. In any case, you end up with a tight coupling between the local DB schema and the service DB schema. This is one of the paths we want to avoid to future proof our apps.

The other approach is to focus on the service requests. The app can queue up the requests and send them when it is reconnected to the service(s). This is typically how e-mail applications work. You do have to handle more work like managing the connection detection, and the queuing and sending of the messages. But it uses less local storage and is a lighter install. Best of all, there is no coupling between the client and the service or data structures on the service side.

In all cases, there are a few common considerations and patterns to detecting and dealing with offline scenarios.

- Asynchronous communications
- Offline cached data
- Timeboxed update history or versioned data
- Local queue of CRUD operations
- Conflict resolution
- Client-side soft deletion of records

We'll discuss these first, then discuss patterns for the scenarios listed above.

Asynchronous Communications: Due to the nature of synchronous communications, where the client makes a request and waits for the reply, it is not suitable for scenarios where the client may not be able to contact a service for an answer. It would constantly time-out when the client can't reach the service. If you write your apps for full asynchronous communications, then the client can send its message to a local queue and go on about its business. This works in online and offline scenarios. The online experience is enhanced because the requests are fully processed almost right away, but the offline experience is not totally broken.

The client should send messages to a local handler that will either pass them on to the service or hold them in a queue until a connection can be made and then send the messages. The once the connection is re-established it can forward the messages to the service for processing. In any case, this is where chunky network communications are preferred over chatty ones. The fewer network calls the app has to make the better chances of success when communication is spotty.

One of the problems when converting legacy apps to modern mobile apps is that the legacy apps tended to be ok with having one service call depend on the result of another. While this isn't a problem on a local network with high reliability, it is very problematic when the client can't connect to the service to get the result to pass to the next call. Applications that will operate in a partially connected scenario need to be designed with fewer calls and where possible eliminate dependencies form one call to the next. This allows the user action to succeed or fail as a whole based on locally cached information without crippling the app when it is offline. If you can design your service calls so that the app can make one call per unit of work, then you will have a better offline experience.

If there is no way to avoid the interdependency of one call on the result of a previous call, then the application needs to gracefully prevent further action, or used assumed or dummy data to complete the dependent call. If your results are fairly predictable, this may be fine, but if you have to guess at the dummy data too much, when the app does finally connect to the service and send the calls, the dependent calls will likely error out due to excessive shift in the actual data from the dummy data. The service needs to be designed to handle these kinds of situations. The service you create needs to be able to handle variances in the calls it receive from the apps, especially where data placeholders were used, and make adjustments for the dummy data and assumptions that dependent calls relied on.

If several calls need to be made to several different services, you should consider having a single service call from the client to a service that can orchestrate the calls to the other services and then correlate a result to be returned to the client. This will prevent the client from hanging in mid transaction or having to abort and roll-back if one of the services cannot be reached or the client goes out of range before all of the responses can be compiled. Sometimes these kinds of components are called Process Managers, Orchestrators, or Controllers.

Offline Cached Data: When planning for offline scenarios, you have to ensure that everything the app needs to operate when it can't connect to your service is available on the device. In the cases of work orders, or field worker apps, that means the dataset with the work orders and other data the user will need while they are out and not connected. Other examples are downloading all the e-mail that has arrived since the last sync. When designing for offline data, plan for stale data and changes made to stale data. Your service has to be aware of the potential for the client to send changes to old or removed records.

You can handle this in one of the three ways. The simplest way to handle it is last in wins. The last update to be sent to the service is the one that gets written to the data layer. While this is simple, it is also dangerous if you have to maintain any kind of transactional history or auditability. This can also lead to chaos when it comes to the users' expectations of saved data. User A may save his changes and assume they are there. When they get back in range and sync, they find out that not only were their changes apparently not saved, but the data they thought they were working with isn't the same as they left with either. This kind of approach is only acceptable in situations where data consistency isn't the primary goal.

Consider the scenario where an escalator maintenance person, Alice goes out on several calls to various buildings around town. She's running late and after her last call she tries to sync over the mobile telephone network but can't. The escalator she was checking is in pretty good shape so she enters the "A-Ok" into the record but can't sync. She doesn't worry about it because she did the inspection and entered the data so she leaves for her holidays.

While she's gone, Bob, the tech covering her territory while she is away, checks up on the escalators and notices one of them is getting worn due to extra load from holiday shopping and needs a new speed regulator before it breaks, so he orders the parts and flags it for replacement as soon as the parts arrive. Everything is synced to the service and the job is marked as awaiting parts. The next day Alice returns, stops in at the office and syncs her app to get work orders. Her old cached records that say everything is A-OK get uploaded, the service writes the records into the data layer because last in wins. The previous record that said the speed regulator needed replaced is over-written with the A-OK record. The parts arrive, but the record shows that they aren't needed so the dutiful intern puts them on the spare parts shelf and life goes on. Over the weekend, a little blue haired lady is injured when her and her hairless cat are shot off the top of the escalator at the mall and land in someone's burger and fries causing the escalator company to reform itself into an amusement park ride company. While last in wins is easy, it's not recommended.

Using forced data expiry is one way where you can stop out of date changes from being applied. The app can keep track of the data's Use By date as set by the service. If the cached data are older than the allowed time span, the app can limit the functionality and require the user reconnect and sync. This is a useful pattern if you want to be sure that decisions are made on relatively current data. It is also useful if you have a highly transactional service where data are changed many times an hour or day and you don't want to have to deal with changes made to data that are several versions older than the current data.

Another way is to allow the client to operate on older cached data and queue up any changes. Once the app is online again and the data are synched, the service will have to handle merging the changes with any changes that have come in from other sources while the app was offline. Usually this is handled with Optimistic Concurrency or Pessimistic Concurrency.

With Pessimistic Concurrency the data are locked for changes at the time the first user requests the data. Subsequent requests for the same data are limited to read-only and the data are reported as locked for editing. Most Source Code Control systems operate this way. Once the initial requestor is finished with the data, the lock is removed and other users can request exclusive access to the data. This is easy for the service and data layer to implement, but it means that only one user can have the data open for update at any one time. This may be fine when the app is dealing with a work order assigned to a single person, or the data being updated are only updateable by a single person such as personal profile information. But if you have many users updating the same data frequently, this is not a very good approach because all users after the first will have their updates rejected. This is what we call "a crap user experience" in the industry.

With Optimistic Concurrency each reader takes their copy of the data and updates it as they see fit. Then when they make changes and attempt to submit them back to the service, the app needs to figure out if it is working on the latest version of the data, and if it is, send its changes. If it is not, it needs to get the latest copy, warn the user of a conflict and ask the user which copy to keep, then submit that copy of the data for saving on the service side. This user experience is better than with Pessimistic Concurrency. But you are also trusting that the user will make the right decision when it comes to saving the changes.

If you have an app where you can't trust the users to make the right decision, and where the data may be changed by more than one source, you need to have a service-side conflict resolution strategy. Unfortunately, there is no silver bullet for this particular problem. Fortunately, it's a fairly rare case in most apps. You will need to build logic into the service or data layer that can make the decision on which version of the data to keep. One strategy is to accept the changes to the most recent known good source of data. This requires data versioning or time stamping.

For example:

DataSet1.V4 is retrieved by Alice. DataSet1.V4 is then retrieved by Bob. Bob makes his change and submits it to the service where it is saved as DataSet1.V5. Charlie then request DataSet1 and gets DataSet1.V5. Alice then submits her changes to DataSet1.V4 and the service rejects them because it's now working on DataSet1.V5 and tells Alice that there is new Data. Alice retrieves DataSet1.V5 and reviews the changes. Charlie submits his changes to DataSet1.V5 and the service accepts them saving them as DataSet1.V6. Alice submits her changes to DataSet1.V5 and the service rejects them because DataSet1 is now on version 6. So Alice gets DataSet1.V6, makes her changes, and submits them where they are finally accepted. Alice gives the app a one star rating alongside Bob and Charlie's 5 star ratings.

While this is frustrating for at least one out of three users in this case, the service doesn't have to deal with merging changes and the resulting merge conflicts, and it mostly solves the problem. However, even this model breaks down under high frequency changes from multiple users. Again, you could let the user decide to go ahead and overwrite the previous changes if your business rules are ok with that.

Many relational databases implement optimistic concurrency internally using one of two techniques that you can use as patterns. This is usually done by way of a time stamp column on data rows. The timestamp of the original row is retrieved by the client along with the data and returned with the update request. The original time stamp is compared to the current time stamp of the row and if it matches then the update succeeds. The other way is for the client to send all of the original data values with the updated data values. The database compares the original values to see if they match what is currently in the database, if they do the data are then updated with the new values and the update succeeds, if the original values do not match then the update fails.

Timeboxed Updates: You want to make sure your app service functionality is prepared for long spans of time between interactions with apps. An app may communicate with your service consistently for a few days, then not communicate again for several weeks. When it makes contact again it might have a significant amount of old requests to send and information to retrieve. This is an extension to the optimistic concurrency method above based on time stamps.

You have to decide based on the nature of your app, if you will timebox those interactions. You may decide that only the past 7 days are relevant to the app. This will require you to allow a rolling window of 7 days' worth of information to be delivered to the client on request. The data structures you use need to be able to deal with one set of data, or 7 days' worth of datasets. Apps that are in constant communication with your service may only need small updates every few minutes, but the app that is only connected once in a while may need a large batch of data sent to it. The data package you send has to accommodate both of those situations.

You will also need either time stamps or most recent data flags to identify clearly how much data to send to the client. You want to be sure that you can avoid sending more data than are necessary. If your service packages up data on a scheduled basis for delivery to the clients, you may be able to use a sequence number for each batch. The client can send the last sequence number it retrieved, and you can then return all data since that point, or up to the maximum allowed by your rolling window.

You may want to degrade the functionality of the app gracefully to only allow operations on the data that are no more than X days old. Actions on data cached on the client that is older than X days can be restricted. Alternatively you can allow operations on all of the cached data and when it syncs again deal with the concurrency issues as mentioned above.

You'll have to decide what is acceptable to your business rules. You may only want to allow additions to data such as comments on social network posts, or you may want to allow changes to existing data such as blogs or Wikipedia entry type systems. The additions portion is easier to deal with. Online social networks operate this way. The timestamp of the comment is based on the time it is received at the service, rather than the time it was created on the client. You may submit a comment about a news story of a blue haired lady staring a new burger and fries hairstyle thinking you're the first one to say anything only to find after you submit your comment that you are the fourth one in the list when the screen refreshes.

Another type of cached data is static or reference data. While this data aren't updated by the client on the app itself, it is frequently referred to and needs to be updated occasionally. For example, if your app records orders of blue wigs sold and you need to add a new shade to the list of options, the new Cornflower Blue color needs to be added to the app's copy of the reference data. The travelling Blue Wig salesman may be just taking orders and not really need to get new data regularly because he's just submitting new order records. So his app may only sync on demand and may not know there is a new Cornflower Blue option available. Your service will have to be able to notify the device that it needs to get a new copy of the color option list. Push notifications are very well suited to this kind of thing.

When reference data are updated on the service side, it should trigger a push notification to the app. The notification can be a simple command that the app interprets to phone home and get the new reference data. You can delay the retrieval until the next sync by setting a flag on the app or have the app try immediately (after all if it received the push notification, it's online).

If you are entirely on a pull model and never initiate communication to the client app, then you need to include the reference data the next time the client app syncs. This approach will potentially delay the update of the reference data, but it is more bandwidth friendly.

Local Queue of User Operations: In partially connected scenarios your app needs to be able to keep a list of all of the interactions the user has done so it can send them when it reconnects to the service. This is commonly called store and forward. The operations the user performed or requested need to be saved locally and sent, usually in order, to the service when possible. This first in first out message sending is ideal for a message queue. You can implement a local message queue in the app and put the service requests into the queue. Once the app detects that it is online again, it sends all of the messages in the queue.

An alternative to this is to save the data or requests as files. Once the app is online again it uploads the files to a location on the service side where the service is monitoring. This would be the case in a photo capturing app or audio note taking app. When the app reconnects, the photos in the "to be synced" folder are uploaded to the service for processing.

In either case, the queue or the file upload, your app and service need a way to track which items have been successfully processed and which ones haven't. If your app is halfway through processing the stored data and the connection drops, it has to know if it needs to retry the one that was being uploaded when the connection dropped, and where to start the next time the app is online again. It's a good idea for the service to send some kind of acknowledgement back to the client when it has confirmed receipt of the queue messages or files being uploaded.

In many cases, most queuing systems you implement will have this functionality built-in. In the case of file uploads though, your app will have to track that even if you use FTP/SFTP you'll have to track the acknowledgements from the ftp transfer.

I would recommend that you do not re-invent the wheel here. You should use existing queuing and file transfer libraries. Most platforms have native or third party libraries you can use to implement message queuing, file transfer, or file sync to your services. Your choice will have to be guided by not only if you are transferring messages which are more suitable to queuing, or files which are an obviously file transfer choice, but security, locking, and standardized protocols as well.

If you use technologies such as message queues over Windows Communication Framework (WCF) you will have to decide if you want message level or transport level encryption, what strength of encryption. You might want to ensure you only sync over SSL/TLS encrypted connections. Every organization should have a set of security best practices that will guide this decision. If you don't have any, take the time to create some or read any of the plethora of books on app dev security.

Data Consistency/Conflict Resolution: Internal office squabbles are out of scope for this book, but we do need to mention data conflict resolution. Conflict resolution is an inevitable part of data consistency if you have multiple users working on the same data. In the cloud-based service world, it is inevitable that multiple users will at some point access the same data. If your application is a single user type application where the data the user touches are only touched by that user, then you won't run into this much. However, if you are writing mobile app services for a multi-user environment, chances are this will matter to you.

I've mentioned data consistency before when I was discussing offline and partially connected devices. I talked about how you can use data versioning to help avoid consistency issues. We're going to discuss this topic in a bit more detail here.

Ideally every environment would implement Strong Consistency. This is the desirable state where every transaction is atomic, if any of the constituent parts of the transaction fail the entire transaction is rolled back and there is always a consistent real-time view of the correct data across all aspects of the system. This doesn't happen by magic though. There are a lot of checks and balances introduced in order to achieve this. Often there are two phase commits in place and blocking service calls that wait for data to complete replication to other servers before continuing. In a high-speed modular cloud-based distributed system this simply isn't going to work very well.

When your database replica may be in a different geographical location, your blocking call may add several seconds to the user experience. In a cloud-based platform, systems that tightly integrate their back-end services in order to ensure this level of consistency aren't going to be right next to each other in a dedicated data center on dedicated network infrastructure. If one of the backup data stores is offline for a reboot, maintenance or disaster, and your initiating service is waiting for the commit from that backup data store, it may have to wait until service is restored. This is one of the tradeoffs when using cloud computing or other forms of distributed computing.

With cloud computing you get massive scale, availability, distribution and impressive cost savings. These are things you can take advantage of to improve your products, and get ahead of the competition. So we have to deal with the issues around trying to implement Strong Consistency. You should only implement full Strong Consistency when you absolutely have to, and there are reasons to do it. Any time you are implementing it though, and you come across a situation where the systems are geographically dispersed, you probably want to reconsider requiring Strong Consistency, and evaluate the platform or alternatives for the result.

More often than not, what you can do is implement data replication at the back-end system. This can often be done asynchronously and uses built-in replication mechanisms in the database, or georeplicated storage provided by the cloud service provider. If you limit the scope of the strongly consistent transaction to a layer above the data layer, this will still fill the need. You can also limit the scope of your Strong Consistency approach to the only data that really needs it such as bank transfers, payments, and things of that nature. Conversely, you can have the Strong Consistency only at the data layer. This is pretty much a standard feature of most relational database systems and can be implemented with NoSQL using Quorums as described in Data Access for Highly-Scalable Solutions: using SQL, NoSQL, and Polyglot Persistence [30].

Keep in mind that even with on-premises systems, implementing Strong Consistency isn't cheap. There is a lot of overhead and a performance hit to doing it. If you can restrict it to only where it's really necessary you may find you can get some performance and complexity gains out of existing systems too. Strong Consistency essentially means that the systems come to a halt until all servers, data nodes, and systems involved in the process acknowledge receipt and safe storage of the information. Updating a change of address or even placing a customer order can probably accept a two- or three-second delay in the data consistency. When you really dig into it you'll find that most things don't require Strong Consistency. In most cases you can get by with Eventual Consistency.

Eventual Consistency is where the systems will all end up with the same data, but it may not be instant and things won't stop to wait for the process. This extends to the point of having geographically distributed systems getting the update a bit later than systems that received it. The update flows throughout the system and eventually everyone gets a copy of the data. If a bank transfer is an example of Strong Consistency where the transaction hasn't happened until all parties agree they have the data secured and the same, Eventual Consistency is the Internet DNS system. When a new domain is registered say www.yourdomain.com, a record is put into the name servers of the company you registered it with. That URL is then mapped to those name servers. That mapping is then copied to all of the routing DNS servers across the internet. It can take up to 76 hours for this replication to complete around the world but eventually, they will all have the record that tells a web browser that (www.yourdomain.com) points to a particular server on the Internet.

While the name server example is extreme, it illustrates the point. Your systems will likely initiate that replication on a much more immediate basis, and you won't have near the servers to replicate the data to that the Internet does.

If you look into the reasons for Eventual Consistency, you will run into this theorem called the CAP Theorem [31]. In a nutshell, it says that distributed systems can't provide all three guarantees of:

- Consistency (all nodes see the same data at the same time)
- Availability (a guarantee that every request receives a response about whether it succeeded or failed)
- Partition tolerance (the system continues to operate despite arbitrary message loss or failure of part of the system)

You get to pick two of the three in most cases. Although the originator of the idea Eric Brewer says that this is a bit misleading [32]. That being said most people accept this and use it as the basis, or even the reason for implementing Eventual Consistency.

If you apply the CAP Theorem directly, it means you can either have perfectly consistent data across the system which blocks all callers until all writes are consistent across all nodes or you cannot block callers but you run the risk of data being inconsistent if some reads are performed from outlying servers before the data are replicated.

Most database systems take the former approach which is essentially Strong Consistency, while most distributed highly available systems implement Eventual Consistency.

Eventual Consistency is quite acceptable in all but the most stringent cases. As I mentioned earlier though, you do need to be conscious of cached data and the issues it will have in an eventually consistent system. You do need to accommodate situations where the delay in consistency can cause problems. For example, if you are booking tickets to a concert, and your browser says there are three left, there's no problem because you only want one. So you take your time choosing your seat. But Bob the roadie is buying for his crew, and buys all three remaining seats while you are checking the seat map. You then submit your choice and are told it didn't work because the seats are no longer available.

The system can handle this a couple ways. It can either just tell you that you're out of luck and show you an ad for the next concert or it can offer to, or even automatically, put you on the standby list for cancelled tickets. I think in most cases, users would prefer the latter. This happens a lot with retail systems. If the item is out of stock by the time you click "Buy" they put it on back order for you. Some of them even operate on the presumption that it's ok to back order everything so it does that first, then sees if there is any stock available.

If your system involves any calls to third party services, then Eventual Consistency is likely your destiny. It's possible that third party services or any service that is external to the current system can be out of contact as shown in Figure 2.3. In any case this kind of coordination needs to be managed carefully.

What happens if only one of the external systems can't be reached? What if it was reached, but there was no acknowledgement back telling the caller that the

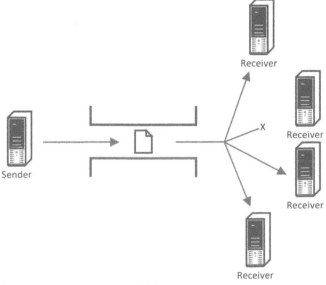

Figure 2.3   System Unreachable

message was received? You need to implement mechanisms that can roll back the transactions if required, or corrective actions can be taken to bring an inconsistent system back into line.

As part of Command Query Responsibility Separation (CQRS), you deal in messages to accept commands for the system to execute. Keeping these command messages allows you to enter corrective messages into the system. You shouldn't replace messages, or change messages, you need to enter a corrective transaction the same way you do in double entry accounting. If you remove a seat at the concert from the list of available seats, and then later the payment fails, you don't just undo the seat reservation. In this case you enter a command to add a cancelled seat to the list of available seats. This is called Compensating Logic.

Compensation Logic essentially defines what to do in the case of an error caused by a duplication of data, error in a message processing operation, or some other problem encountered by the system. It is especially useful in long running multi-step workflows or processes. It is a way of correcting errors should one of the steps fail. In some cases it is also a convenient mechanism for reversing operations at the request of the caller as well. For example, if you decided you wanted to cancel your concert tickets, the compensating transactions can be used to process that request.

In large-scale distributed systems some of the downstream, external, or third party systems may have already committed the change to the data before the error occurs. This error has to be corrected but it can't just be rolled back because some systems have already completed the commitment process. This is not the end of the world. You can implement a Compensating Transaction pattern [33] to correct the system and re-establish consistency. The compensating Transaction reverses the effect of the erroneous transaction. This takes a bit of planning to implement. Ideally a retry will work and save you having to run through the compensation logic, but in case it doesn't here's how the pattern works.

- **Track what each command does and record the compensation logic for the step.** This constitutes recording what would undo whatever the operation is doing. Every command being executed in the system should generate compensation logic for itself. For example if you are booking your concert seats, canceling that seat is the compensating logic. The operation and recording its compensating logic need to be atomic. You don't want to have compensation logic waiting to be run if the original command didn't complete in the first place and you don't want to have to compensate for a completed operation without the compensation logic being recorded.

- **Detect failures of the operations or steps in the operations.** Monitor for failures through whatever mechanisms make sense for your scenario. You may be using asynchronous callbacks that indicate operation success, or you might be using synchronous calls in some places that return values you can check for exceptions. Once failures of an operation are detected, initiate the compensating transactions.

- **Roll back the transactions.** Execute the compensation logic for each step (this may not need to be done in reverse sequential order, you might be able to execute the compensation logic in parallel). Ensure that this is done in accordance with any applicable business rules. For example, if you cancel your concert tickets, you may not be entitled to a full refund.

Keep in mind that failure of one component does not necessarily mean failure of the entire series of operations. If you wanted to book your concert tickets, including a back stage pass and a limo, just because there are no back stage passes left doesn't mean you have to reverse the tickets and limo. Ideally you'd be given the choice of a Mosh Pit Pass instead or free beer.

In the CQRS pattern, Event Sourcing acts as a natural tracking of operations. Implementing compensating logic would be a matter of mirroring those command messages and inverting their operations. Later we discuss the CQRS pattern in more detail as a way of implementing a highly scalable service.

As you can see, this is not trivial, but it does work very well. There is a lot of work that goes into implementing this. As with all things software, the more work the developer's do, the less work the administrators and users have to do. Implementing retry logic is a much easier path to take and is supported by most database and messaging infrastructure systems.

Retry Strategy: Instead of dealing with compensation logic, you might want to implement a retry strategy. If the initial call to a system fails, you need to retry the operation. There could be many reasons for failure in a large-scale distributed system ranging from temporary outages to system load and service updates. The easiest thing to do is to have the system retry operations that failed for system level problems. These are things like systems being unreachable, offline, etc. Obviously a retry doesn't make sense when the failure was due to business rule violations such as bad user input, or bad data in a file or a kitten trying to purchase its own catnip. A retry would be pointless in those scenarios.

With retry systems, you need to either make sure that if a failed operation is entirely rolled back leaving the system in a previously consistent state, or you need to make sure all of your operations are idempotent. That way if you do re-run an operation that did succeed, but the calling system didn't know it succeeded, the re-run of the operation won't change the system by doubling the number of seats reserved for example.

While this sounds easy enough, it's actually a bit tricky. So much so that there is an established pattern called an Idempotent Receiver pattern [34]. While you can implement duplicate message handling in the message passing infrastructure, it is a non-zero performance hit. It is easier to create an idempotent receiver and let the messaging infrastructure pass messages as quickly as possible. If you are dealing with multiple third party systems, and one of them fails, it's much easier for the system to just send the message to all receivers again knowing that they are idempotent and they can deal with the duplicate on their end.

You can accomplish this in two ways. The receiver can explicitly ignore duplicate messages, or the messages themselves can be designed to be more absolute than interpretive. For example, instead of saying "Remove 1 Seat from the Availability List" the message says "Set the Available Seats to 4" This is pretty clear and easy and doesn't take a lot of code on either end, but it's also somewhat dangerous. With this approach you have to be very careful about ensuring message order and consistency. This would work fine in a Strongly Consistent scenario, not so much in an Eventually Consistent scenario.

When you design the receivers themselves to be idempotent, you have much more control and you can still operate with Eventual Consistency. You do this by giving the messages some unique identifier. This can be anything that makes sense to your system but is unique for every message of the same type. In some cases this will be a database primary key, unique user identifier such as e-mail address, or even GUIDs if you want to use something that is referenced from somewhere else.

An alternative to using assigned identifiers is to use uniquely generated ones. This is actually not has hard has it sounds. Essentially, you use a hash of the message. You can create a hash (MD5 for speed or SHA256 if you are paranoid) of the message and use that as your key as in Figure 2.4. It would work something like this:

- Receiver pops message off of queue
- Receiver calculates MD5 hash of message
- Receiver checks list of processed messages to see if it contains this hash
- If not, process message else drop message and delete it from the queue

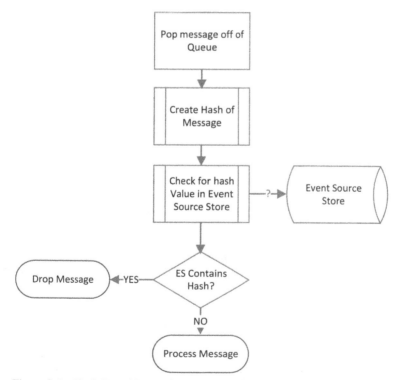

Figure 2.4   Hash-Based Dupe Message Detection

If you do use assigned identifiers, in a pinch they can be reused if there are enough unique ones that they are not going to be reused within the timeframe that a transaction being run twice would break the data consistency. For example any system that is keeping score of the most tweets in a 24-hour period only needs enough identifiers for all of the tweets (plus a margin) that the system will process in 24 hours. This will allow the use of int values, for example, as opposed to GUIDs which are larger and less network friendly.

As each receiver receives and processes the message, it records the unique identifier for that message. You can even record the entire message if you need it later for auditing and compliance purposes which are often done in CQRS/ES systems. When processing messages the receiver checks to see if this ID is in its processed messages list and if not it processes the message, if it is, it simply ignores it as having already been processed.

For consistency and extensibility you should also declare your operations as idempotent so that other systems connecting to yours can be confident in resending messages as their retry strategy. This will help not only with the integration of your system with third party or external systems, but it helps future proof your system by making things very easy to adopt with little ambiguity.

Client-Side Soft Deletion of Records: You may be in a situation where you need to keep records of data or what has happened to that data for a long time. You may even have to be able to recreate a system state from years ago for audit, legal, or regulatory purposes. Allowing clients to permanently delete data doesn't serve this environment very well. You need to keep records that have been deleted in the offline cached data and eventually in the online data store as well.

The best way to do this is through Soft Delete. Soft Delete is essentially using an "active" flag on the data so as not to actually remove the data from the system. Each record in the data store has a Boolean flag on it called "Active." When the data are live and operational, that flag is set to True. When the client has the data on the mobile device, when they delete a record, that Active flag is set to False. All queries and views of data filter by that Active flag. Data without its Active flag set are not sent to the client when syncing or requesting new data.

This effectively removes the data from the active system without permanently deleting it. At any point in time you can examine the data as they were when they were "deleted." You can go back in the record history and reconstruct the context for auditing and other similar events. This is also easier and faster than removing an entire record.

2.2.7.2 The Recommended Approach to Each Online Category
Now we'll look at how we'd approach each category of online app. Each one needs slightly different models and patterns to get working smoothly. We aren't worried too much about the Online Required and Online Not Required models because they are pretty straight forward. It's the partially connected scenarios that we need to discuss.

Online Preferred: This scenario is one where you expect the app to be online most of the time. It provides the best user experience when online. Some of the features may require being online but the app is functional offline. Weather apps and News Readers are good examples of this category. A weather app will download the forecasts when it is online, but if you disconnect and go back to the app, you can still see the last forecasts that it retrieved. But if those forecasts change, you won't get the updates and won't know that a sudden snowstorm will hit in summer. While the app is connected, every time you go to the Forecast screen it checks for fresh data and gets the latest up to the minute forecast. Obviously being online is preferred for this kind of app, but it's not required.

Dealing with this kind of app is a matter of detecting the connection state, if online then check for stale data, then retrieve it if necessary and using small amounts of local storage on the device cache it locally. Almost every device now allows apps to use a small amount of local isolated storage that is provided just for that user and that app. The latest data should be packaged and time-stamped with an appropriate expiry time. When the user goes to the screen with the data, the app checks the timestamp, then polls the service for new data if required. An alternative to this approach is to have the service either push data to the app or push a refresh notification with the most current timestamp. Then the app can compare the timestamps with what it has and fetch the new data if required.

This prevents the app from polling all the time and using up bandwidth and battery. Remember polling from the device is the last resort for automated updates. Pushing from the service is preferred because it is more device friendly.

Update patterns:

- On startup check for online status and retrieve fresh data if any
- App polls for new data when the local data expires
- Services push new data to app
- Services push a New Data notification with current timestamp, app compares times and requests new data if necessary

Online Regularly: In most cases this approach is similar to the Online Preferred scenario. However, now we have to consider how regularly the app is online and how we handle stale data. You will have more data caching on the device in an online regularly scenario. These kinds of apps will sync on a fairly predictable basis such as hourly, daily, weekly, monthly but are expected to be offline just as regularly. You'll take two approaches to the data expiry. You will either not expire it or you will expire it after the sync period is overdue.

Cases where you might not need to expire the data are ones where the user of the app is the only one working on that set of data. If it is an insurance adjuster or a maintenance person who has a specific set of clients or equipment that is assigned only to them, data expiry isn't as important. This is because it is unlikely to change while they have the data and their app is offline.

If, however, you are working with data that are not exclusively used by a single user, and there is a chance that another user may modify the data after the first user has gone offline, you will need to consider expiring the local cache of data to force the user to sync. This ensures that if there are changes, the user's data are not so far off of current that it causes problems. It is that timespan that dictates when you need to enforce the regular sync. Some organizations have even gone through the process of determining a percentage of data change threshold.

For example, for every X transactions that happen to the data, there is a Y percentage chance that any arbitrary user will be affected by that change. When the Y percentage is over a predetermined threshold, say 30%, force the sync. Then they figure out based on their transaction rate, how many hours, days, or weeks can pass before that 30% mark is hit. They set their data expiry to that duration.

You know that these apps will be connected on a regular basis and because of this, having the user initiate a manual sync is fine. While you can do it based on connection state, this may end up syncing the data more often than is necessary which is not bandwidth friendly.

There are occasions when the system knows that data have changed for a particular user that was not initiated by that user. In these cases you may need to prompt the app for an Out Of Schedule data sync. You can do this usually through a push notification to the app if the app is online and sync is done manually. If the app is completely disconnected for the full duration of the sync window, then the data will have to be refreshed and conflicts should be handled client-side before the clients' changes are synced back to the system.

Update patterns:

- App primarily uses cached data.
- User initiates sync when online.
- Data expiry set to regular interval based on business rules or likelihood of data change.
- Service Pushes an Out Of Schedule notification to remind the user or force the app to sync when a known relevant change occurs.

Online Optional: With Online Optional apps, the app is actually not expected to be online in any predictable manner. This is a pretty common scenario in games, for example. They are only online to get updates or new content. But they are designed for offline use in most cases.

This kind of scenario means that you don't have to worry about data expiry, forcing sync, or pushing data changed notifications to the client. But you may want them to go online to get new updates for the app, or to purchase in-app items which are only available from the online store. There are two approaches to doing this. Nagging with local reminders and nagging with push notifications.

When nagging, you need to decide which kind of nagging your users will be the most tolerant of. There are many games out now that regularly pop-up messages to the user reminding them that there are new features available. If the app is offline, then you need to have a background process of some type that fires off a pop-up reminder to the user once in a while. Just a friendly reminder that their minions need them and there are new minion hairstyles available in the online store. This can all be done on the client, but it's not easy to refresh with new data.

The other form is through push notifications. These allow you to send limited information to the user that is actually fresh from your services. These are really good for letting the user know that their friends just beat them at the latest cyber caber toss and that they need to upload their scores. The key though is that there should be no degradation in user experience, if the app never goes online again.

Update patterns:

- App operates completely offline
- Data update entirely at the user's discretion
- No data expiry
- Users can be requested to go online for new or additional data and in-game items but the experience should remain the same if they chose not to

2.3 WRAPPING UP

In this chapter, we've covered a lot of ground in platform independence and offline scenarios. One of the keys is to be kind to the device. Make sure that you are kind to the battery and bandwidth. Move what processing power you can to a service back-end. Save your on device processing for what needs to give the user the optimal experience.

Due to the mobile nature of these devices we have to plan for offline scenarios. Most of this has to do with managing data caching and synchronization. It is difficult to ensure Strong Consistency in these cases so you'll need to get comfortable with Eventual Consistency. If you plan for that upfront, your decisions on how you handle data and turn chatty conversations into chunky ones will be better off. You'll also be able to plan a good data versioning and reconciliation strategy upfront where you need to get the architectural patterns ironed out.

Considering the various offline scenarios and if you are going to be faced with Online Preferred, Regularly, or Never, will help you decide what kinds of restrictions you'll need on the data caching. You might want to plan your stale data policy early so you can build in the correct handling mechanisms for it. In later chapters, we'll see some of the mechanisms you can use to do this.

CHAPTER 3

PLATFORM-INDEPENDENT DEVELOPMENT STRATEGY

3.1 HIGH-LEVEL APP DEVELOPMENT FLOW

We have established that to get instant breadth an HTML5 website is a great way to go. However, to get a richer and more personalized experience a native app is required. At this point, we will look at a strategy to achieve both while not painting ourselves into the single platform corner.

To achieve breadth and depth, we need to start from a base that will give us the best reuse and portability of our core code. We need to be able to remove as much duplication between platforms as possible while being able to create rich native app experiences. To do this, we will start from a breadth standpoint to give us the most visibility possible while we prepare our native depth-focused apps for release on the various marketplaces.

In general, the process starts from an HTML5 website, then progresses down a path of least resistance to deliver native app user interfaces with minimal rework on the various platforms. Figure 3.1 helps describe the apps as they are released and how they derive from each other.

This process helps you get the most reuse out of your code. In cases where you use Xamarin as your primary codebase, you might need to convert the UI to Xamarin.Forms so that you can compile the UI down to native iOS and Android. You can even use it to compile for Windows. Those kinds of things won't work for Web though, so you'll still have the HTML5 codebase. To get as much mileage out of that as possible, you might consider PhoneGap/Cordova so you can use the same code and skillset to build your native apps. The general idea is to start with a codebase, that you can get the most reuse out of, while still being able to produce native apps.

Shared class libraries and portable code are critical to this process. Build as much of your UI layer processing into shared libraries as possible. Limit the platform-specific code as much as you can. This reduces your code proliferation which makes bug fixes and updates orders of magnitude easier.

Designing Platform Independent Mobile Apps and Services, First Edition. Rocky Heckman.
© 2016 the IEEE Computer Society, Inc. Published 2016 by John Wiley & Sons, Inc.

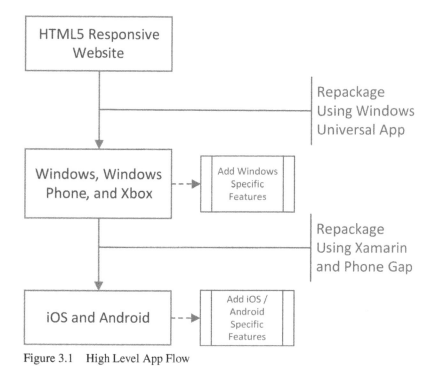

Figure 3.1   High Level App Flow

3.2 FIVE-LAYER ARCHITECTURE

Traditionally architects have designed an n-tier architecture. In most cases, n was 2 or 3 tiers. In client-side apps, they were often one layer monolithic apps. In modern business apps we have usually required three distinct layers. Traditionally, according to Fowler [35] these were called the Presentation Layer, the Domain Layer, and the Data Source Layer. As applications have become more device focused, and the services they talk to tend to be remote and usually reachable via the Internet, the traditional patterns get changed a bit. In this architecture, we have the User Interface (UI) layer, the Business Logic (BL) layer, and the Data layer. This serves most purposes and works well with most enterprise applications. Consumer applications often only had one physical client layer for an app. More often than not, this resulted in one, maybe two, logical layers as well. Usually client-side apps were logically broken into a UI and a Data layer, if the architects and developers were forward thinking. In the new platform-independent delivery world we use five distinct layers. By using five layers as shown in Figure 3.2, we get our future proofing.

The three primary layers, UI, Service, and Data layers, are what we have had in traditional client-server architectures. The UI layer handled the display of data and collection of user input. The Service Logic layer often referred to as a Business Logic layer handled the manipulation of data, number crunching and was generally the brains of the system. The Data layer is where the data were physically stored. In several cases business logic was built into the Data layer by means of complex stored

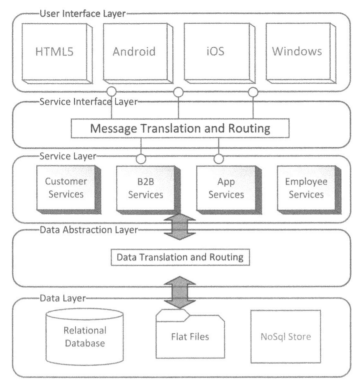

Figure 3.2   Five Layer Architecture

procedures. While this is fine for data manipulation and integrity checks, too often complex business calculations and workflows were implemented at this level. These layers in large part still perform these roles. However, to eliminate the temptation to build these logical layers too closely together in physical layers, or to have our database serve as a workflow loan calculator, we have added the two interconnection layers.

The Service Interface Layer and the Data Abstraction Layer, are what provide us the future proofing and agility we need to adapt to the changing mobile computing landscape. They provide distinct decoupling between the app layers. They also force us to follow strict guidelines for service presentation and data handling which helps to maintain a consistent and predictable set of interfaces. We will go through these layers in detail to explain their purpose and cover some guidelines to ensure they operate as smoothly as possible.

3.3 FIVE-LAYER ARCHITECTURE DETAIL

3.3.1 The User Interface Layer

The top layer is the UI layer. This is the thinnest possible layer of code that displays data from the services and interacts with the local device. To ensure the most flexibility and future proofing, the code in this layer needs to be only what is required for a rich user experience and no more. Traditional mobile apps have put almost everything in this layer which prevents any code reuse, restricts rapid platform changes, and limits the developers' ability to respond to user feedback quickly.

This also results in massive codebase proliferation. Every time a new platform comes out, because everything has traditionally been written into the one device layer, the entire codebase had to be re-written to port the app to a new platform. Part of the problem was that a lot of the mobile platforms were intentionally designed to use very different languages and systems to create apps for them. The case that drives the point home is the iPhone.

In the start of mobile applications when business ran apps on Windows Mobile devices and Palm Pilots, there was a fairly well-established system of using Visual Basic or Visual C++ to create apps for these devices. But Apple didn't want developers to be able to create apps that ran on the iPhone that could also run on another platform. When iOS came out on the iPhone, apps had to written completely differently even though the language was similar, you couldn't take existing C++ apps and convert them to Objective C to run on the iPhone very easily. Not to mention hardly anyone used Macs when the iPhone first came out. So Apple forced people to buy a Mac because they don't allow you to submit an app to their App Store form anything else. This was terrible for developers because everything had to change completely, but it was a brilliant lock-in play for Apple. To avoid that problem in the future, we need to keep the code that gets deployed to the device as minimal as possible.

While modern devices are getting more and more powerful, people are also expecting more out of them. Since we want to keep the mobile device code as minimal as possible, we don't want to put too much application logic in there if we can help it. Remember, for most apps, what is on the device is purely a presentation and data input layer. It's not meant to be a full-fledged desktop client application. So we need to be disciplined and keep the more complex logic and data processing in the Services layer.

The other reason to keep processing and sensitive logic off of the device is that once it's in the users' hands, you can't trust it anymore. If a user has access to the code, then it is no longer your trusted code. This has been a weakness of many thick client apps and some poorly written web apps. They assume that whatever is on the client is safe, so often developers overlook security and data validation on the server. There is also the added problem of intellectual property leakage. If the user with your app on their device wants to see how you did something, with the right tools they can crack open the app and have a look. So if you have some super special proprietary algorithm, try to keep that off of the device.

The less code we have written and deployed that is specific to the devices, the more agile we are, and the less coded proliferation we have. This means fewer codebases to maintain and fewer locations to fix bugs if they arise. The rule of thumb for the UI layer:

> *Rule of Thumb: Keep the UI layer code as minimal as possible while still providing a rich experience.*

3.3.2 The Service Interface Layer

The Service Interface Layer (SIL) is a message routing and translation layer. Its primary job is to decouple the client-side apps from the physical services and it's one of the most important abstraction layers for future proofing your apps and services. This interface layer allows developers to move services around from on-premises to the cloud, or go from hosting their own service, to utilizing a partner's service and routing the traffic to it instead of the service they were hosting. It also allows a developer to create new versions of the service and host them side by side or seamlessly upgrade existing users to new services. Figure 3.3 shows how this works.

By keeping the client-side endpoint the same and using the Service Interface Layer to route traffic to a different service, we future proof our client-side because we don't have to deploy new client apps when service locations or versions change. If we think about the services behind the SIL, we have a lot more freedom with this structure.

We can even take entire sections out of our services layer, and move them to another location, or hand that functionality off to a third party. We can then route the traffic to the new location without disrupting our client applications. If required we can even implement a shim in our service layer that can translate from what our clients send and expect to the format the third party needs. This is a huge step toward a solid future proof design.

Additionally, when we consider security and the abundance of Distributed Denial of Service (DDoS) attacks on infrastructure, having a robust SIL that can rebuff the DDoS attacks adds stability and reliability to our products. This SIL is also your first port of call for being able to perform all of your data validation and authentication.

It is where you will do the message translation for incoming and outgoing messages. It is where all of your endpoints will be published for apps and other services

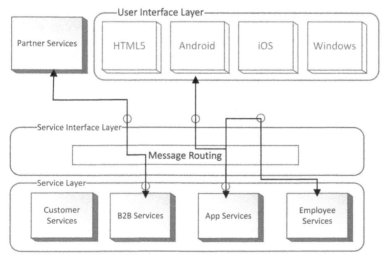

Figure 3.3   Service Interface Layer Routing

to contact your services. This layer is what provides us the future proofing for our systems and allows us to quickly adapt to changes in the mobile landscape.

3.3.3 The Service Layer

The Service Layer is the workhorse of our app offering. It performs all of the calculations, workflow, authentication, data manipulation, and is the core of the services of the future. This is where your value-add is. In fact, several large ISVs have started giving away the client-side app for free with basic demonstration functionality. They then charge subscription fees to access the services they maintain to provide the backend functionality for the apps to use real data, more services, and enable additional features. This is where the bulk of your effort will go.

This layer can be built using whatever technology you want. As long as its performant, and you understand it or you will spend a lot of time learning and trying to find help. That being said, if it is time to learn new skills then do so but plan this into your project and its deployment schedule. As long as the technology can handle the inputs from the Service Interface Layer, and send and retrieve data from the Data Abstraction Layer, then it will work.

You will also want to consider what kind of technologies will work in a large scale, or at least expandable environment. Many of these services will be deployed in cloud-based computing environments. So they will need to function that way. The biggest issue facing many systems that migrate from on-premises to cloud is that they are not designed for high availability and failover. If you have the opportunity to create new Greenfield systems, then they can be designed from the ground up to support things like high availability and failover. This is really just typical web farm design and deployment like enterprises have been using for high-availability systems and service-oriented systems for quite a while.

With a high-availability design, you want to ensure you avoid things like in-memory session state, local file writes, and hard-coded configuration. The system may have to failover from one machine to another in the case of a disaster and it has to be able to do so without losing information pertinent to the current user sessions. When you use in-memory session state, for example, if the last machine to receive a user request get rebooted, or Godzilla steps on the data center, then the user is going to lose whatever they were doing. Take this in the context of the app though.

If your app does not have long multi-page forms, or information that needs to carry from one session to the next, you might not have to worry about an external session state server. You might be able to just keep temporary session information in memory. But anything like a shopping cart or a high-score table, and especially multi-page form data or customer information will need to be designed for an external session state server. Alternatively you can use some form of distributed caching mechanism as long as it's external to the service and its server. Redis cache has been a very popular choice for these kinds of implementations. More information on the Redis cache can be found here: http://redis.io

This layer will also need to call services that take care of cross-cutting concerns such as authentication and authorization, logging, and auditing. These kinds of systems are often implemented at this layer, but should be separate from the actual application service logic itself. You don't want to re-write your authentication code each time you create a new service. This quickly becomes a problem, especially if you have a bug in the code and you have to fix and re-deploy all of your services to fix the bug. We will cover these issues in more detail later when we discuss highly scalable service architectures.

3.3.4 The Data Abstraction Layer

The Data Abstraction Layer is similar to the Service Interface Layer in that it provides a decoupling layer between the services and the physical data storage. Its primary role is to translate between code readable objects and physical data representations. This kind of layer can usually be implemented with most Object Relational Mapping systems. For example, systems like Hibernate/NHibernate, Entity Framework, and others translate between code level objects and back-end data stores such as relational databases, NoSQL databases, and flat files structures such as XML files. The implementation system itself acts as the DAL. However, you can choose to not use these systems and create an abstraction layer of your own. This usually will involve some kind of serialization system, XMl, text, binary, or other, to files or into some kind of relational data store.

ORMs also provide some of the future proofing attributes that we want to build into our systems. If we are building our app services based on XML file storage, and the app grows in popularity the users might want features that would require a relational data store such as an Oracle database. In this case, we can migrate to those new physical data storage systems and simply re-route our data abstraction components such as Hibernate or Entity Framework to target the new destination data store. The service layer doesn't have to change its understanding of the data. Although in truth a change this major would likely mean you want to add features and will probably enhance the services anyway, so it's a good thing we have our Service Interface Layer. Overall though, the less the Service Logic Layer knows about the physical data storage mechanism, the more future proofed you are.

You will also need to factor in message routing to external services that you may use as a data store. This will usually be done by routing the traffic up through the SIL. You may even have to work with a polyglot data system where you have multiple different kinds of data stores. In cases like these you may have to implement a query processor or a repository to deal with assembling the data from the various data stores into something your services layer can work with in a contextual way.

The DAL is also critical to future proofing the system against changes in the underlying data storage mechanisms. You need to get this layer done correctly to ensure you don't do yourself harm in the long run.

3.3.5 The Data Layer

This is the physical storage mechanism the services persist data to. These are normally relational databases, flat file storage, or unstructured storage such as NoSQL databases. The storage mechanisms here can also be other services from partners or service providers. For example, you may have some of your own physical Oracle databases, and you may have a payment clearing house service that you send payments to, as well as using a cloud-based Software as a Service (SaaS) database such as Amazon Dynamo DB or Microsoft Azure SQL Database to store data shared with partners or other services. All of these represent data stores that would logically reside in the Data layer.

Your application shouldn't have to care where or how the data are physically stored. So building in an intimate knowledge of the physical data storage mechanism will break the decoupled design and prevent you from easily changing data storage systems in the future. This is why all access to the Data layer needs to go through the DAL. Changes to the physical data layer should not affect the Service layer.

Data storage technology changes pretty regularly and often what seems like a good idea now may prove to not be a good idea when the app grows to 5× or 10× in scale. A good example is using something like MySQL for a database. MySQL is a great relational database. It's versatile, simple to use, and most importantly free. Many content management systems such as WordPress use it and use it well. However, it doesn't scale well and it doesn't do many things that enterprise grade databases do such as Oracle and SQL Server. One of the key benefits we get out of more enterprise grade databases is the ability to do analytics research on the data we've collected.

In a modern app, with hopefully millions of users, you will want to get some trend analysis, demographic data and be able to predict where we want to put your improvement efforts. This will help you in the future extending existing apps and designing new ones.

This doesn't mean you don't start small and affordable with something like MySQL or Raven DB, but it does mean that you have to plan for scale and upgrading. Having everything in a dedicated Data layer, with a data abstraction layer over the top gives us that flexibility. It means that some of our apps can use MySQL while others use Oracle, and yet others use flat files. You can upgrade the back-end database or change it from one vendor to the next without having to redeploy the entire stack. While the Service layer is your value add, the Data layer is where you collect on your investment.

CHAPTER 4

THE USER INTERFACE LAYER

4.1 PORTING VERSUS WRAPPING

When we think of how to get our applications from one platform to another, there are two methods that come to mind. Porting the code over which involves writing the code again in the language native to the target platform, and wrapping the code in something that provides an interpretation layer. In the cases of updating legacy systems to be mobile apps, we often have to change language, and indeed the base architecture of the system to fit the highly distributed service-oriented nature of mobile apps. Ideally, if we are starting fresh, we would write the application in a language and system that provides the porting or wrapping for us.

Porting is converting code and then compiling it down to the native bytes for the target platform. Xamarin compiles your C# code down to the right native bits for the target platform. Wrapping is putting an interpretive layer around your code and packing it up with an interpreter that actually runs your app on the device. Unity is a game development platform that wraps your code. However with wrapping, your code won't run at all unless that interpreter layer is there. In most cases, porting is preferable to wrapping due to producing a more native experience.

Again we are presented with the Breadth versus Depth dilemma. Porting gives you depth but can result in more work and more bits of platform-specific code to maintain. Wrapping provides more breadth, but you are in a "least common denominator" mode and won't be able to take advantage of platform-specific features. Fortunately most of the platforms you would write for, have a pretty common feature set. Unfortunately, they are rarely the same from a code perspective.

If your target market is a breadth one such as a game or social interaction system, wrapping may be the best choice. You can write the game in Unity which provides the Unity Player for all major platforms including the Web. For other mobile apps you may want to write them in Apache Cordova or Adobe Air which can provide that interpretation layer for your code to run on the supported platforms. In either case, the runtime interpreter has to be present on the target device, usually it's packaged with the app itself. The catch is that you have to either abandon platform-specific features that are not generic enough to work cross-platform, or use third party SDK's bundled into the deployment that you can call to access the native features.

I've often heard "Oh no, it requires you to download something to run it like Flash" and people get upset. Well, sometimes that is fine. It may not work when there

---

Designing Platform Independent Mobile Apps and Services, First Edition. Rocky Heckman.
© 2016 the IEEE Computer Society, Inc. Published 2016 by John Wiley & Sons, Inc.

are restrictions on what people can download and install which is typical of most commercial environments, but it's not that uncommon in consumer environments. At some point in the past few years, most people have already downloaded Flash Player, Adobe Air, Java, and more recently the Unity Player so they will already have it and won't need to download it again. While I have to admit to personally not favoring this approach, it serves its purpose in the right context.

I prefer the approach of lowering the barrier to entry as much as possible. This means reducing the dependency on external components. If they have to download something, in addition to your app, it may be a step too far for them to bother. If the interpreter or runtime can be embedded with your app package, that's a better approach. The more third party things you have to package with your app also means the download and install is larger. Streamlined apps and app packages is the best approach. Eliminate as much from the app package as possible, without sacrificing user experience or rich app capabilities. For example, you will likely include third party UI controls and components for charting, etc. but you don't need to include debug libraries and things you use for testing.

If your target market is more of a depth play, which is typically the case with line of business applications, or apps targeting at a specific demographic, you may want to use a porting system to generate native apps. Tools such as Xamarin help with this angle. You write your app with familiar technologies such as XAML/C# and Xamarin compiles it down to genuine native code for the target platform. For example, with Xamarin, if you target Android, it produces a Java based app, if you target iOS it produces an Objective C/Cocoa based app. In both cases they are native to their target platforms and don't require an interpreter layer or separate runtime to work. It's easier to access and code for native features this way because the native constructs are normally available to you.

4.2 MULTI-CLIENT DEVELOPMENT TOOLS

4.2.1 PhoneGap (http://phonegap.com/)

PhoneGap from Adobe is based on Apache Cordova. Applications written in Cordova, and hence PhoneGap, are written using HTML5, JavaScript, and CSS3. This may sound familiar to anyone who wrote applications for the early Windows Phone 7 or Windows 8 platform. Modern apps written for those platforms were able to use the same set of technologies. Applications in this technology set got a bad rap for being too simplistic and lacking depth. However, this is the developers' fault. As shown above with applications like the web-based version of Microsoft Office, or the World's Biggest Pac Man feature rich, and attractive apps are very possible in this technology set, if you take the time to do it right.

Applications in Cordova are really single-page web applications. Some or all of the application is packaged in a WebView. In some cases, the WebView is part of a larger application that may be written in the platforms native app system. For Cordova based apps the main screen is even named index.html. The apps are wrapped in a native application wrapper and are presented as a WebView within that wrapper as a stand-alone app, or embedded into larger more complex apps.

This is pretty good if you want apps or pieces of apps that can be reused across various platforms. Where this starts to get complicated is when you want to interact with the native hardware or take advantage of platform-specific features. Phone-Gap/Cordova based applications require plugins to communicate with the native platform components. For example, if you want to communicate with the Compass on an Android or Windows platform, you need the plugin that allows you to call the native compass from the JavaScript in your Cordova based application. You can't expect to communicate with a compass on all devices because some of them don't have them or don't expose an API for them such as the iPhone 3G or Blackberry. So you have to rely on a plugin for the platforms that do have it. Cross-platform common capabilities such as Network or Camera that work pretty much the same across devices will have native support in PhoneGap.

Once the app is built and wrapped in the appropriate platform wrapper, it can be submitted to the various app marketplaces for the platforms. Most of the compilation and packaging is done from the Cordova command line. As of this writing Microsoft is introducing Cordova support into Visual Studio 2013/2015, there is a phonegap-gui project on GitHub (https://github.com/hermwong/phonegap-gui), and XCode (http://www.adobe.com/devnet/html5/articles/getting-started-with-phonegap-in-xcode-for-ios.html) as well as Eclipse for Android development (http://www.adobe.com/devnet/html5/articles/getting-started-with-phonegap-in-eclipse-for-android.html).

Originally the drawback to PhoneGap/Cordova based apps was that they all looked the same. This means you don't get the native UI experience that you would expect on the platform. Whereas iOS has tabs on the bottom, Android has them on the top. But the Cordova app would have them on the sample place on either platform. This may be the effect you want, especially if you want to allow your users to transfer from one device to the next and have a familiar UI. But if you want to blend with the platform you are on instead, you might need to look at Microsoft Universal Apps with the Xamarin plugin or Xamarin itself. However now Cordova allows you to customize the UI for the target platform the same way Xamarin and Visual Studio do.

4.2.2 Xamarin (http://xamarin.com/)

Xamarin is a cross-platform mobile app development system based on Mono, and now on the Open Source .NET Core. Not surprisingly it is a cross-platform development system where the code is written in C#. It then compiles that C# down to languages that are native to the target platforms, so the code is compiled down to Java Intermediate Language for Android which is just in time compiled on the device when the application launches. The apps are packaged into an .apk file. For iOS the applications are compiled to ARM assembly code and packaged into a .app file. From Xamarin's own web page: (http://developer.xamarin.com/guides/cross-platform/getting_started/introduction_to_mobile_development/)

> *"Xamarin is unique in this space by offering a single language – C#, class library, and runtime that works across all three mobile platforms of iOS, Android, and Windows Phone (Windows Phone's indigenous language is already C#), while still compiling*

native (non-interpreted) applications that are performant enough even for demanding games.

Each of these platforms has a different feature set and each varies in its ability to write native applications – that is, applications that compile down to native code and that interop fluently with the underlying Java subsystem. For example, some platforms only allow you to build apps in HTML and JavaScript, whereas some are very low-level and only allow C/C++ code. Some platforms don't even utilize the native control toolkit.

Xamarin is unique in that it combines all of the power of the indigenous platforms and adds a number of powerful features of its own..." [36]

Xamarin also uses a runtime to handle direct device interop memory management and garbage collection.

Xamarin allows you to share core code between the various app types through project types in the solution. You have Portable Class Libraries which are consumed as binaries or Share Application Projects which are added to the solution and compiled with the app project. This lets you take advantage of as much shared code as possible, while keeping the platform-specific UI code in separate projects.

Xamarin lets you build applications in their own stand-alone Xamarin Studio, or use Microsoft Visual Studio if you are already familiar with that environment. In either case, you can produce cross-platforms apps from either editing environment. Keep in mind that while you can deploy to the Windows and Android marketplaces from either Mac or PC, you can only deploy to the iOS app store from a Mac. That may have an effect on which environment you chose to develop from.

The bottom line is that anything you can do in Java or Objective-C to deploy to Android and iOS respectively you can write in C# with Visual Studio and Xamarin. So you don't have to maintain multiple codebases and multiple skill sets. This is a massive savings in development effort. The apps you write your C# code for are native on their target platform and look native with native library usage for the platform-specific hooks like camera, GPS, and storage.

4.2.3 Unity (http://www.unity3d.com)

If games are more your style, or you want highly visual commercial apps and simulations, you may want to look at Unity 3D as a cross-platform development environment. Unity is a combination of a game engine, some assets, and a development environment. At the time of this writing, according to Unity 3D themselves [37], games based on Unity were holding 45% of the market share and the closest competitor 17%. Unity supports 17 different platforms including PCs, phones, and consoles.

Unity is actually a game engine that you write apps to run on. This means that anything you write using Unity will require the Unity Player to be installed on the target device for your app to run. Fortunately, the Unity Player is quite widely distributed already as several popular games and apps have already been written with it and anyone that has played those games or used those apps already has the Unity Player installed on their device. In fact, the Unity Player is already integrated into the Qihoo 360 Safe browser used extensively in China so that they don't have to download the player for Unity based games to work.

Unity games are developed in the Unity editor and can be written in C#, Boo, or JavaScript. Boo is a somewhat proprietary language that Unity games were originally written with. Unity has started to lean toward the C# language though as the primary language for developing in Unity. Most of their first party Software Development Kits (SDKs), like their Facebook SDK, are written for C# development as is most of their documentation.

4.2.4 Visual Studio

I want to mention Visual Studio here specifically. Microsoft Visual Studio is the most widely used Integrated Development Environment (IDE) in the industry. It is also arguably the most comprehensive and best IDE in use today. This makes it the one you will likely use in anything other than pure iOS or pure Android development. With things like XCode, you can only develop for iOS, and Eclipse is great for Java, but you can't use it to develop for iOS or Windows Phone/Tablet.

Fortunately, all of the current cross-platform development systems such as Xamarin and Unity have first class plugins for Visual Studio. What this means is that the best cross-platform app development tool is Visual Studio/Xamarin, and the best cross-platform games development tool is Visual Studio/Unity. With Visual Studio 2015, the cross-platform integration is even better with the introduction of iOS emulator capability and remote debugging of iOS. The only other thing you need is an Apple Mac in a corner to do the compilation and deploy the app to the Apple Store with. You can also plug in your iPhone to the Mac and do native remote debugging on the actual iPhone through Visual Studio.

Visual Studio has also integrated Apache Cordova directly into the IDE. So Cordova based development is a first class citizen in Visual Studio. Visual Studio is emerging as the premier cross-platform development IDE with help from plugins like Xamarin and Unity. The Community edition of Visual Studio is also free. In early 2016 Microsoft acquired Xamarin and it is now included for free in Visual Studio.

4.3 CROSS-PLATFORM LANGUAGES

This section is not so much a list of cross-platform development languages as it is a discussion about the approach and what you should consider when choosing one. In reality where it matters is on the client side. When you write the UI layer for the app itself you need to use whatever language is supported on the device. The three top runners are Java, Objective-C, and C#. While Apple is making the move to Swift, Java and C# are making the move to true cross-platform. When Microsoft released the C# compiler which you can even get yourself here: https://github.com/dotnet/roslyn and the server side .NET Framework to the community via open source [38], it made C# a very viable language across all platforms. Java has already enjoyed that status for some time. Objective-C and Swift are not cross-platform, but if you want an app that runs on iOS you need to use it.

What's more important is to limit the amount of codebases you have wherever possible. As I mentioned above, if you use something like Xamarin and its native controls, it will produce native Java, Objective-C, and C# binaries to be deployed to the target devices. But if you want to use some other development environment, you will need to keep track of the individual UI codebases separately.

When it comes to the back-end services, the answer is, use whatever you want to. The app, or partner services, or third party apps won't care what you wrote it in as long as it accepts the service calls and returns the expected results in a format they can understand. In reality, for you to be productive and do your best work, you need to work in the tools and languages you are most comfortable with that will get the job done.

There will forever be a religious debate on what language is best. In some cases like iOS that choices are taken out of your hands, in others like Windows and Android you do have a choice. Ideally, you should take a cross-platform approach from the very beginning. So if you can use a tool like Xamarin that can do the UI layer code translation for you, then do it. If you have multiple teams that each specialize in a particular platform, let them specialize. But just make sure your services are where you put the bulk of your effort and you can use whatever you want at that end.

4.4 AVOID WRITING FOR THE LEAST COMMON DENOMINATOR

One of the things that users tend to notice is when an app doesn't "feel right" You'll get their attention right away if all of a sudden your iOS or Windows app looks like a web page out of the 1990's and it's running on new sleek hardware. But why would you do this in the first place? The answer is it's the shortest way to cross-platform UI. HTML5, in all its glory, still can't quite compete with native apps. They are more responsive and more importantly they use native controls which make them look and feel like the apps that come from the first party developers. You want to be part of that device experience. It's very tempting to do to just slap your HTML5 website in a web frame and submit it though.

The nirvana of writing an HTML5 responsive web app and then being able to port that to a native app with things like Apache Cordova seems easy. What's more interesting is that it is. Using PhoneGap or Cordova in Visual Studio means you can generate a native, store downloadable app with one set of HTML5 code. But, the app will look exactly the same on every device. This is great for a consistent user experience, but bad for a device and user experience point of view. You will not get favorable reviews with what looks like a web page on a phone unless you take the time to create UI screens for each platform.

Consider menu location placement. In Windows the ellipsis to open up the app menu is at the bottom right, in iOS the hamburger button for the menu is at the upper right having recently been moved from the upper left, in Android it's on the upper right. But in a Cordova based app, it's in the same place and does not necessarily match the standard for the device. If you are calling native functions for things such as dialog boxes, then it will leave it up to the device OS, but UI elements you design into the HTML5 will look the same across platforms by default.

It's a bit of a challenge to get a true native look and feel across platforms when the manufacturers themselves are determined to make sure they don't do things the same way. But I suppose if they did they'd sue each other into oblivion. In any case, if you are going to use an HTML5 style UI, take advantage of the CSS capabilities to adapt the UI and style of the app to the device. You can build your apps with different folders for each platform and put the respective UI screens in there. Then most tools like PhoneGap and Xamarin or Visual Studio will use the appropriate UI screens form the folder for the app platform it is building for.

When it comes to Cordova, I think this quote from MSDN magazine sums it up best:

"Cordova's Sweet Spot: Developers generally realize that running HTML, CSS and JavaScript inside a Cordova app won't always produce a truly native experience on every platform. Cordova apps, then, are ideal for highly branded experiences that needn't conform to native UI patterns, and for enterprise apps where differentiation across platforms isn't desired. Cordova is also an efficient means to quickly build pro-duction apps that test ideas in the marketplace, helping you understand where to make deeper business investments." [39]

4.5 WRAPPING UP

The UI layer is the one that everyone will judge your app by. It is the sexy bit that people see. You want it to look and feel native, and offer a great user experience. This got harder when different platforms started following different design guidelines.

With cross-platform tools like PhoneGap, Visual Studio, and Xamarin, you can create apps that have a common code layer, but different UI screens for each platform from a single codebase. You'll want to decide on your approach, either wrapping or porting up front.

If reuse is your highest priority, then you want to use something like PhoneGap or Cordova and reuse as much of your HTML5 website code as possible. If you want a better native integration experience you'll want to go with Xamarin and build your UI in Xamarin.Forms.

Whichever you choose, take the time to build screens for each platform that look and feel native. Users won't like apps with a jarring UI that doesn't fit the device it's on as much as they'll like ones that feel like an extension of their device. Be part of their lifestyle!

CHAPTER 5

THE SERVICE INTERFACE LAYER

5.1 MESSAGE PROCESSING

There are several things to consider when you think about what you want success to look like for your apps. Hopefully success is your app on millions of devices, or for a corporate app, being the go to advantage for your organization. In either case, you will have to deal with lots of data quickly. You will also have to consider if the clients will pull messages from your system or you will push the messages to them.

In most modern mobile systems the client is not always online, and they may only use your app in bursts. For example, many people will use SMS on their phones. They will type out their message and hit send. The phone is usually back into their pocket before the message has even left the phone. Users don't tend to wait for things to happen. As a society our expectations have been set to align with our attention spans. According to popular thought, an app should allow a user to open it and do something productive in the time it takes a barista to make your coffee.

This means that people don't want to be waiting around while your app talks back and forth to the services. The traditional method of synchronous communications is not going to align very well with our modern coffee-based schedules. Mobile app services should be based on asynchronous communications. That is to say, use asynchronous calls by default and synchronous only when you must.

While most modern programming languages can do this now, platforms like .NET are using async calls throughout most of the framework. You will find the C# async attribute and await keyword all over C# code now, and that's a very good thing. If the language you are using is not treating asynchronous communications as a first class citizen, you need to reconsider your platform. Fortunately the languages used for the major platforms all have, or can emulate, asynchronous method calling. I can't see this going backwards in the near future so I presume new platforms will also support it.

With that in mind, one of the ways to implement asynchronous communications between clients and services, service to service, and even different layers of the services is through message queues. Message queues are a long-standing well-understood mechanism for communicating between systems in traditional Service-Oriented Architectures (SOA). Since most modern mobile apps are very much like clients consuming services from an SOA, it makes sense that the paradigm would transfer.

Designing Platform Independent Mobile Apps and Services, First Edition. Rocky Heckman.
© 2016 the IEEE Computer Society, Inc. Published 2016 by John Wiley & Sons, Inc.

Fortunately in modern systems message queues can be accessed over common web-based protocols such as HTTP, REST, SOAP, and more traditional WS-∗ based web services. They offer many advantages such as guaranteed delivery, message retry, and even transport and message level encryption if required. Most cloud platforms either support running traditional message queuing platforms such as WebSphere in virtual machines, or have PaaS-based services that you just sign up for and use such as Microsoft Azure Service Bus. In either case, you'll be able to use it for your service architecture.

That being said, there's nothing stopping you from direct calls to SOAP- or REST-based web services. You can call them directly and even call them synchronously if required. Remember the golden rule. In many cases in fact calls from the client to the service will be standard web service or WebAPI calls. Ensure that whatever you use will fit the purpose you need it for.

5.1.1 Push versus Pull

There are two ways you are going to get messages from one end to the other of the communications channel. The sender can push them to the receiver, or the receiver can pull them from the sender. Determining if a system is a Push or a Pull is decided from the perspective of the receiver of the message. Receivers will either pull them, or the messaging system will push them to the receivers. This is an important decision in mobile app and service design but fortunately there are key factors that will basically decide for you in most cases.

5.1.1.1 Push Systems   In a Push system, the sender creates a message and puts it onto the messaging system. The messaging system then takes over and sends a message to each receiver that has signed up to get the messages. Think of the way social media apps like Facebook and Twitter pop up a little message when you get a Facebook notification or Twitter direct message. You weren't actively using the app, and it's not sitting there asking the sender if there is a message waiting every couple of minutes.

These systems are particularly good and getting data to users without them having to come to you. If you have an app that needs to notify a user of something, or that wants to be able to tell the user something even if they are not actively using your app, you should consider a Push Notification System (PNS).

The drawback to PNS is that they require the messaging infrastructure to keep a list of all of the receivers so that you can send messages to them. It not only has to keep a list, but in apps where messages can be sent to specific people in that list, it also has to handle an addressing mechanism so you can send your message to individual receivers.

The simpler version is more of an Event Notification system. In these systems, messages are put onto the message queue and the messaging infrastructure sends that message to every receiver in the list. This is a bit easier for the messaging system to deal with because it just has to go down the list and send the message to everyone. It eliminates the need for the addressing mechanism. However, that list still has to be maintained.

5.1.1.2 Pull Systems   In a Pull System, the receiver initiates all of the communications with the messaging infrastructure when it comes time to receive the message. The app has to periodically check the messaging system to see if there are any messages waiting for it. While this is easier on back-end senders and the messaging infrastructure, it is harder on the client.

When a client has to poll the messaging infrastructure every time it wants to know if there are messages waiting, it costs battery, and bandwidth. So pull systems are much less device friendly than a push system. That being said, pull systems are much more service friendly.

From the service side you can handle this in two ways. First, you can eliminate a messaging infrastructure all together if you prefer. You can simply set up a web service or web API that the client app calls with its credentials or unique identifier. The service then looks up any notifications or new data for that client and returns it as part of the response. This does mean having an extra mechanism in your service though. If you prefer for your back-end service to be more insulated from the client-side, you can still use messaging infrastructure to do this.

You can set up message queues or channels. Your service can then put messages destined for each client on the particular channel assigned to that client. This means that you can send information to clients, and they can retrieve it at their leisure without touching the service directly. As noted in Section 3.3.2 The Service Interface Layer, this is using the SIL to its fullest. The message request can be passed through the SIL to the Service Layer which can send back the messages destined for the client that initiated the call, or the service can put the messages on a queue in the SIL, or a dedicated service that just handles messaging which is accessible through the SIL.

5.1.2 Partially Connected Scenarios

Another aspect that needs to be considered when deciding push versus pull is partially connected scenarios. These are situations where the sender and receiver are not in constant contact. So think of places where there is spotty mobile internet coverage, basements, mines, and the country of Chad. In these areas push notifications may not work without the messaging infrastructure being able to hold messages until the client pops its head up and it can deliver the message.

For this reason, messaging systems introduce guaranteed delivery. This means that it will hang on to the message until it knows the receiver has gotten it. In a pull system, it just waits until the client comes and gets it. In a push system it keeps trying to send the message until the receiver gets it. This push attempt is usually triggered by an out of band process that detects when the client app is online, and then triggers the push.

In most modern PNS from the major players like Apple, Google, and Microsoft, the system has a monitor to tell the PNS when the client is online and reachable so that the polling problem isn't just shifted to the PNS. If you were to create your own PNS, you would have to build this into your system. If not, your messaging system will spend countless hours pinging for clients. This is a real waste of, well everything.

As we discuss in Section 1.4, it's better to hand of as much as possible to the pre-built PNS systems. Otherwise you'll have to design and build all of the bits that they've already built such as a way to monitor for device connectivity, manage the queuing system for accepting, holding, routing, delivering, and so on, the messages. Again, this is a build versus buy decision, but I think that due to the complexity of PNS systems, if you want to get up and running quickly I'd suggest considering the buy option in this case.

5.2 MESSAGE PROCESSING PATTERNS

We have to decide how our apps will communicate with our services. Do we want them to call the services directly, or use some kind of message passing system? If we are fine with a client submitting information directly to the server, we can just use web services and web APIs. However we want to make sure we can be as versatile as possible which also means asynchronous processing. Web service calls tend to be synchronous, as do most request/response systems. But in order to ensure we can process information, messages, and general communications we need to consider message queues.

Message queues provide a mechanism to pass communications between the app and the service without causing either to wait for the communication to complete. If we have to wait on either end, we tie up resources, network and battery. You will see the intentional implementation of asynchronous communications permeating all aspects of mobile services and architectures.

In this discussion, we will refer to senders and receivers but we need to clarify what we mean here. A sender is anyone or anything that creates the message to be sent and processed. If the client is creating data for you to process, the client is the sender and your system is the receiver. If your system is creating information for the client(s) to process then the system is the sender and the client the receiver. Senders and Receivers are types of apps or processing back-ends. Even though you may have 3 or 4 instances of your back-end process running for high availability, scale, and failover, the process is essentially one sender/receiver. The Processor on the back-end is one logical unit. There may be multiple instances of the receiver for scalability, but there is one intended consumer of the messages. Think of it as a logical unit, not a physical implementation. Each sender and receiver essentially see the message queue as its own channel so from the communications perspective there is one sender and one receiver.

For example, you may have a photo sharing app. Each app on the client is a sender, and your photo processing service in the cloud is a receiver. While there may be millions of users with the app on their phone, we still think of the app as a single sender. Your back-end processor will likely have more than one instance of it running, but it is the same code across instances and is therefore considered a single receiver. Ok, I think that horse has had enough kicking.

Three are a couple of common message passing systems in use today. Most of them were created for the Service Bus style communications used in corporate SOA systems. Message queuing being the most popular. Messages would come in and be

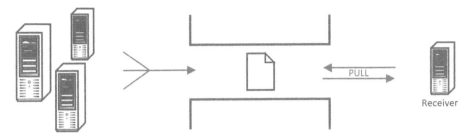

Senders

Figure 5.1   Many to One Pull Message Queue

put onto a First in First out (FIFO) queue. The responsible service or distributer would pull the messages off and process them. This works fine and there are mechanisms for ensuring that a message is only processed once, and that it can be re-tried if it fails, or if the message is unable to be processed it can be put in a holding pen (referred to as a dead letter queue) to be looked at later.

Now we have several versions of this kind of system. There are three types of these systems in common use that you will use in your apps. Note that in this list, where I show a single Sender or Receiver that could represent multiple instances of the same node. Where multiple Senders or Receivers are shown, they are different nodes, not multiple instances of the same node.

- Many to one channel: This is where multiple senders are sending messages to a single receiver. This is the standard web server/service model where multiple clients call the same service endpoint. It's such a common scenario that any web-facing system is assumed to be setup this way for at least the primary channel. In the case where the communications are done via message queues, you have this pattern shown in Figure 5.1.

    Pros: Very standard architecture. Queuing implementation is basic.

    Cons: Due to multiple receiver instances pulling messages from the same queue, you may end up with a Competing Consumer problem. Message queue will have to implement single receiver controls to prevent duplicate processing of messages. If the receiver is the client app, the app must poll for messages.

- Point to point channel [34]: Standard message queues which take a single instance of the message from the sender, and when that message is taken from the queue and processed by the receiver it is gone. The receiver initiates reading the queue and if there is a message waiting it picks up its message. This process is indicated by two arrows on the receiver end in the diagrams in Figure 5.2. This particular pattern is used to ensure that a message goes from a single sender to a single receiver. That does not mean that there is only one instance of the service on the receiving end, but that only one of the instances may pick up and process a message. This way you don't have multiple receivers processing the message multiple times. This is a way to maintain consistency while letting the message queue enforce the single receiver delivery. Once the

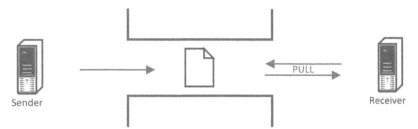

Figure 5.2   Point to Point Channel

message is picked up by one receiver instance, the message queue prevents other receiver instances from picking up the same message.

Pros: Eliminates the Competing Consumer problem. Allows for easy scale out of Sender or Receiver. Guarantees messages are from a single sender to a single receiver.

Cons: More complex queuing system implementation to avoid Competing Consumer problem. Not suitable for Many to One, One to Many, or Many to Many configurations. If the receiver is the client app, the app must poll for messages.

• Publish/subscribe (Pub/Sub) [34]: In this case, a single message is put on a queue, and many receivers have subscribed to the queue. There are multiple different receivers of the message. The queuing system itself keeps track of the subscribers and keeps the message available until all subscribers have gotten a copy. This is how a lot of news and social media systems work. This is typically a One to Many Pull system. It is not a broadcast system because the receivers have to specifically subscribe to the channel and come get the message. Whereas a broadcast system tends to spray the message out onto the network to anyone able to receive it. Receivers don't have to explicitly subscribe to a channel, they just have to be listening.

Pros: Sender can create a single message and reach multiple receivers.

Cons: Message subscribers must be tracked to know when all subscribers have picked up the message. The receiver must poll for new messages or pull them as in Figure 5.3.

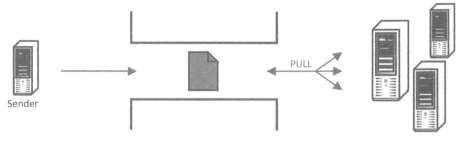

Receivers

Figure 5.3   One to Many Publish/Subscribe Message Queue

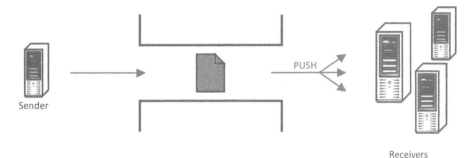

Figure 5.4   One to Many Push Message Queue

- Push notifications: In this pattern, a single message is put onto a distribution channel. The notification system then sends a message to every client that has asked to receive messages sent to the channel. This is a Push system. The messaging handling system itself initiates the communication and sending of the message to the client. This differs from a publish/subscribe system in that it prevents the client from having to poll the queue periodically to see if there are any messages waiting. Figure 5.4 illustrates this.

    Pros: One sender can reach multiple receivers. Receivers do not have to poll

    Cons: Requires specialized or dedicated systems to push messages to different platforms. Data pushed are usually pretty limited in size.

Depending on the type of app you're creating and the type of processing you require, you will decide on which of these you will use. For systems where the apps will be sending you data, you will use the Single Instance Processing model. For apps that aggregate information from several sources, and send data to several clients, you will use the Pub/Sub model. For most apps where you want to reach out to the users and let them know of a data change or new event, you will use a Push Notification model. We will cover the pros and cons and choices in more detail next, but for now here is a quick chart to help you decide.

5.3 HIGH-VOLUME MESSAGING PATTERNS

Now that we have an idea of the basic message processing patterns, what about processing really high-volume messages? In most cases the patterns above will work fine, if you can scale out the messaging infrastructure appropriately. The problem you will typically run into is the limited queue depth on most queuing systems. In many cases, the queue depth is set at a relatively low number. For example, in IBM's WebSphere MQ 8.0 the default queue depth is 5000, with a maximum of 999,999,999 [40]. A billion messages in a queue. While that seems like a big number, it's not. Consider our IoT example from above.

If you have an app on an IoT device in a car, you may receive ten million messages a second. That means you have about 99.999999 seconds before the queue is full. Your system would have to be able to process ten million messages a second just to keep up and prevent the queue from reaching its maximum depth. If your system hiccups or slows down and can only process 9,999,999 messages a second, you will lose the race and the queue will fill up. Fortunately, there are systems for handling this kind of volume.

In most cases enterprise message queue systems from vendors such as IBM, Oracle, Microsoft, are designed to handle this with parallel queuing. However, you will need a significant investment in these products and infrastructure to handle heavy load. Cloud-based message handling infrastructure is far more economical and scalable.

5.3.1 Queue Services and Microsoft Azure Event Hubs

Microsoft Azure Service Bus(SB) Queues and Amazon's Simple Queue Service (SQS) work in a similar manner and are focused on scale-out traditional queuing. SQS claims the ultimate in scale out with unlimited senders/receivers, messages at any time. [41] This kind of commodity service is becoming commonplace. It takes traditional queuing system features including long polling message batching, and of course dead letter queues. Cloud providers make them scale out with consumption being based on a pay for what you use model.

In order to provide this kind of scale, something has to give. In this case, the SB and SQS messages are limited to 256 KB of text in any arbitrary format. Billing for SQS is done per each 64 KB chunk of data so one 256 KB message is worth 4 "requests" on your bill. SB bases the pricing on API calls to the queue instead of the number of messages sent or received. [42]

Something else that Amazon offers is using SQS in conjunction with their Simple Notification Service (SNS). This allows you to send one message to the SNS, and it will be able to send it to many SQS queues. This can be a handy way to do broadcasting to queues, or multiple polling receivers that do not have the ability to receive push notifications.

The MS Azure Service Bus as you might imagine is a cloud-based messaging infrastructure designed specifically for large-scale messaging. Microsoft Azure Service Bus is a hosted, highly available infrastructure for broad communication, massive scale event distribution, and service endpoint publishing. You can connect to Service Bus through Windows Communication Foundation (WCF) and REST endpoints that would otherwise be hidden behind a firewall or hard to route access to. Your service endpoints can be located behind network address translation (NAT) firewalls, or on dynamically assigned IP addresses, or both and service bus can still help you get traffic to them.

Service bus has relayed and brokered messaging functionality. Service bus relays can do direct one-way, request/response, and peer-to-peer messaging. The brokered messaging functions provide durable asynchronous messaging through Queues, Topics, and Subscriptions, which is great for decoupling and using pub/sub patterns. [43] This supports our partially connected scenarios very well because messages to and from the client can wait on the queue until either side is connected and ready to process them. The service bus is the message highway for apps that are built in Microsoft's Azure platform.

SB relays [44] are a very useful feature for getting messages back to on premises services that you can't move to the cloud. They allow you to put a public endpoint in Azure, and the SB Relay will securely route the traffic back over a VPN to your service sitting in your data center. This allows you to put a buffer between you and the big bad internet. You can close down all the ports in your firewall (except the VPN) and not expose your data center directly to the internet. Everything will only see the public endpoints on the SB. This is a really good use of things like service bus for organizations that are more security conscious. We discuss this deployment pattern in Section 1.5.2.1.

Like most cloud-based queueing systems, SB and SQS can be monitored by the auto scaling components to know when to add more receivers to the cluster when the queue depth reaches a certain point. This is a very useful feature that cloud providers offer to help you keep your costs down and only pay for what you need, when you need it. This prevents you from having to spin up a bunch of receivers to handle the load "just in case" or provision infrastructure for peak periods that sits idle for most of the year.

There are times though when you need massively high-speed scale. This is especially the case in Internet of Things, or embedded sensor nets, or anything else where you'll need to accept hundreds of messages a second. Traditional queueing isn't suited for this due to the high overhead involved in managing the queues. A different kind of message transport is needed in these cases. The current standout cloud service for high-volume data ingestion is the Microsoft Azure Service Bus – Event Hubs service. Event hubs are a subset of the MS Azure Service Bus.

Event hubs are a feature of the service bus that is designed specifically for high-speed, high-volume message ingestion. Because of this, it does not support all of the features of a traditional message queuing service such as dead letter queues, sequencing, and transaction support. If your massive mobile app user base, or IoT app generates massive messages to your cloud-based mobile app service, then you need to consider event hubs. If you do need the more handshake and dead letter management capabilities, look at other technologies such as Service Bus Topics and Queues, or similar products you can install and run like MSMQ and WebSphere MQ.

One of the problems you encounter with standard queuing systems is that all of the receivers read from the same topic or queue. This can create a bottle neck at the queue level since it can only respond to one queue read request at a time. Even though you have all of the instances of your node wanting to read the same queue and get whatever message is next, there is still that problem of the queue being able to hand out one message at a time. There are a couple of versions of this.

- Receivers pick up whatever message is there and process it
- Receivers pick up only messages addressed to them

An analogy for the addressed messages scenario would be picking up badges at a massive conference, think Consumer Electronics Show, Electronic Entertainment Expo, GenCon, or your favorite massive conference (if you don't have one, just go with it for a minute). All of the conference badges are being produced and are preprinted ready for collection. When you arrive there is only one volunteer up front handing out the badges, and forty thousand of you waiting in line to collect them. The badges happen to put into a chute that only lets you take one out at a time from the end and there is no guarantee they are in any kind of order. One of the happy conference goers steps up to the volunteer and says Hi, I'm Mac Byte can I have my badge please? The volunteer looks at the chute and reads the name. He looks back to Mac Byte and informs him "Sorry, that's not the next one, come back in a minute." Conference goers keep stepping up and asking for their badge until the person whose badge is actually next in the chute takes it and the process starts over again. This is a massive processing problem picking up messages.

It's only slightly better in the non-addressed version. All of the conference goers still have to wait for that one volunteer to hand out one badge at a time. This is a standard queue or topic scenario. You need to use partitioned queuing to get around that problem.

With partitioned queuing there are multiple chutes and a volunteer at each chute handing out badges. If it is an addressed message scenario, each chute will have the first letter of the last name above it and all of the badges in that chute will be for people whose last name starts with that letter. This allows us to break up the conference goers into 26 separate lines and we can hand out 26 badges at a time. Alternatively if the badges aren't named specifically, then we can use a round robin distribution and attendees can go to whichever line is shortest at the time.

Event hubs are designed from the ground up for partitioned queuing to accommodate the really high speeds and volumes. Each receiver is pointed at anyone of the partitions for the messages. There are many partitions that all have messages being added to them, and that can all hand out messages at the same time.

For example, if we have a game service that receives player updates from thousands of players a second, we can add those messages to the event hub, and it will spread them across the appropriate partition for the message either based on a supplied partition key for the player, or game they are in, or in a round robin fashion if no partition key is supplied. Our services connect to the event hub through the AMQP 1.0 session to receive the messages. They are placed into a group that is targeted at a specific partition key, or a generic consumer group for non-partition specific message retrieval.

Event hubs, in order to achieve the massive throughput automatically assign messages a time to live. This means we don't have to worry about queue depth. Event hubs can store billions of messages so messages are dropped out of the partition based on time, rather than stopping message ingestion like a queue does when the queue depth is reached.

If you have multiple systems that want to process the messages from event hubs, they can. Event hubs won't kill the message once it's been read until the time limit is reached. So you can specify time-based offset to read your messages from. If the message counting service is processing messages really fast, while the message processing service is taking a bit longer, the counting service can specify a more recent offset to pick up messages from which brings it closer to the newer messages. Using two pointers for this as in Figure 5.5 makes this more efficient.

If you just want to process all of the messages in sequence and all of the services in the consumer group are processing the messages, you can use Checkpointing.

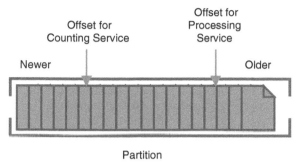

Figure 5.5   Event Hub Offsets

With Checkpointing each of the services sets a checkpoint of the last message it read from the partition. Then the next service to request a message you can have it start processing after the last checkpoint. This is how you can effectively mark messages as having been processed. If something goes wrong or you want to look at older data you just specify the offset or the checkpoint at which you want to pull messages from.

5.3.2 Web Sockets

There are occasions when the apps or web UI will need to be more synchronous, or at least appear so. In these cases typical HTTP request/response won't work as it's too slow. If your app is more interactive like a game or real-time collaboration tool, you will need a persistent connection or a way to make it look like it.

   You want to be able to do this without multiple TCP connections to each client, the client being forced to continually poll the server for new information, and the overheads incurred from the HTTP request/response cycles for each call. You want to be able to use the Universal Firewall Traversal Ports, 80 and 443 without arguing with security over new holes in the firewall. This is where Web Sockets comes into play.

   Web Sockets are like regular network sockets but one layer up and can be programmed from JavaScript on the client-side or various libraries for whatever platform and language you are writing in. They establish a persistent two-way communications channel between the client and the server. Messages can be sent both directions without establishing a new connection or more importantly packaging and sending a new request and spinning while waiting for a response. This gives the server side the opportunity to initiate message to the client as well because WebSocket is full duplex.

   As stated in RFC6455 [45]:

Conceptually, WebSocket is really just a layer on top of TCP that does the following:

- adds a web origin-based security model for browsers
- adds an addressing and protocol naming mechanism to support multiple services on one port and multiple host names on one IP address

- layers a framing mechanism on top of TCP to get back to the IP packet mechanism that TCP is built on, but without length limits
- includes an additional closing handshake in-band that is designed to work in the presence of proxies and other intermediaries

Other than that, WebSocket adds nothing. Basically it is intended to be as close to just exposing raw TCP to script as possible, given the constraints of the Web.

WebSocket connections are established through an HTTP Upgrade request. The client sends a request similar to the following:

```
GET /chat HTTP/1.1
Host: server.example.com
Upgrade: websocket
Connection: Upgrade
Sec-WebSocket-Key: dGhlIHNhbXBsZSBub25jZQ==
Sec-WebSocket-Protocol: ticker, trader
Sec-WebSocket-Version: 13
Origin: http://example.com
```

The server responds like this:

```
HTTP/1.1 101 Switching Protocols
Upgrade: websocket
Connection: Upgrade
Sec-WebSocket-Accept: s3pPLMBiTxaQ9kYGzzhZRbK+xOo=
Sec-WebSocket-Protocol: ticker
```

Just a point of note. The Sec-WebSocket-Key: dGhlIHNhbXBsZS Bub25jZQ== in the client request and the Sec-WebSocket-Accept: s3pPLMBi TxaQ9kYGzzhZRbK+xOo= in the server response are quite specific. The server must take the value of the Sec-WebSocket-Key: header, and concatenate it with the GUID 258EAFA5-E914-47DA-95CA-C5AB0DC85B11 then generate the base 64 encoded SHA-1 hash for that value and return it in the Sec-WebSocket-Accept: header. This mechanism is designed to ensure that there is some level of validation for the WebSocket handshake. For more information on the headers and what they mean, see the IETF RFC 6455 [45]

The sub-protocols can be anything you want as long as your client and server agree on them. The client sends a list of protocols it can or wants to use. The server responds with the list of those protocols out of the client list that it agrees to use. The server does not respond with all of the protocols it can use, but only the ones it can use out of the list that the client asked for. This prevents bleeding too much information to clients that are fishing for ways to attack the system.

If you create application level protocols, you should register them with the IANA registry to make sure you don't end up with protocol collisions, and you don't have to re-invent the wheel if a good one already exists that fits your needs. Ideally you're going to register your protocol with your domain name. For example myprotocol.mydomain.com. This also allows you to version them like: v1.myprotocol.mydomain.com and v2.myprotocol.mydomain.com. This is important when it comes to creating B2B and B2C services. Customers and business partners will need to be able to speak your service language, and these protocols are how you define that language.

Once the connection is established, the client and server can send messages to each other in a direct non request/response manner. You could just send updates down to the client without the client talking to the server again. You send your message according to your protocol over the WebSocket connection and act on those messages accordingly.

5.4 HIGH-VOLUME PUSH NOTIFICATIONS

When you are dealing with lots of mobile apps, you may need to send notifications to thousands or millions of devices at a time. Consider apps like the ones used for tracking the results of the Olympics. These apps sent notifications to millions of people when medal tallies were posted, or localized notifications when countries people registered interest in were competing in selected events. How do you handle this kind of mass push of information?

Just to provide some context, how do you do this in a cross-platform world where each ecosystem will have its own notification system. For example, Windows Notification Service (WNS), Microsoft Push Notification Service (MPNS), Apple Push Notification Service (APNS), Google Cloud Messaging (GCM), Amazon Device Messaging (ADM) and Baidu Cloud Push for Android in China (Baidu). This becomes one of the problems that takes some planning. If you want to do cross-platform notifications, you might want to consider existing third party services to do this. If not, you can certainly roll your own, but be prepared for a lot of updates and changes and leading platforms change, and the providers change their platforms for new features or to keep them proprietary.

This becomes a build versus buy decision. If you want to build something like this, you won't want all of the services you have that need to send notifications to be forced into knowing about multiple proprietary notification systems. You'll need to build a Translate and Forward Service (TFS) that sits in the Service Interface Layer.

The TFS is responsible for taking push messages from the services layer, translating them to the appropriate proprietary format, assigning the appropriate channel authentication credentials, and then forwarding the translated message to the proprietary Push Notification Services. You want to centralize this kind of component because if something changes in one of the PNS systems, you can make that change once to this component and all of your services in the Service Layer can continue to operate with no impact. If a new platform comes along, and a new PNS service

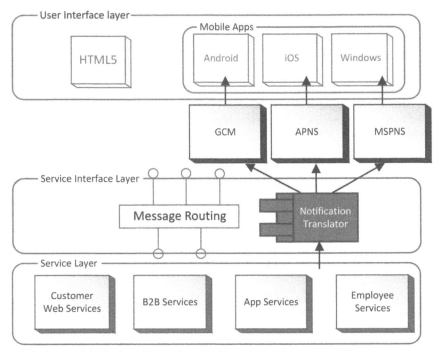

Figure 5.6   Multiple Proprietary Push Notification Services

is created, you just add that to your TFS and you can push notifications to the new platform with minimal development.

I'm sure you can see that if you were to put this level of code into each individual service, a change to one of the PNS systems, means you have to change every service that has the code to send to the affected PNS. Additionally, as shown in Figure 5.6, it's better to have this kind of component in the Service Interface Layer, rather than in the Service Layer. This is because it is essentially a message routing component. It should not contain business logic, does not call back-end components, and provides an abstraction layer between external PNS systems and your services.

Because this component is a one-way outbound message pusher, it can be optimized for outbound traffic. Firewall rules and security rules can be applied or minimized for outbound traffic only. The component will not have to accept any inbound traffic and could effectively be closed off from any such traffic. It still has to be able to communicate outbound, but not inbound. The Service Interface Layer normally has its external-facing endpoints in the demilitarized zone (DMZ) and it's internal-facing endpoints in the internal network.

If you do build your own version of this, you will have very good control, but it does mean more code to manage. It also means that you will have to maintain another self-owned component on your cloud provider. There are first party cloud solutions for this problem as well as several third party products that perform this kind of work. If you'd rather pay for this as a service, or a licensed component, it means less work for you in most cases.

5.4.1 Third Party Notification Hubs

Sometimes it's easier to buy the service than build it. Both Amazon and Microsoft have solid offerings for Push Notification Services. They both support most common programming languages .NET, Node.js, Java, PHP, Python, and Ruby. Both support developing for Android, iOS, and Windows devices. Microsoft additionally specifically supports Flash and Silverlight. So chances are you'll be able to target most things from these platforms. You can download the SDKs from their respective locations Amazon: http://aws.amazon.com/tools/ and Microsoft: http://azure.microsoft.com/downloads/.

Amazon provides a service for push notifications called Amazon Simple Notification Service (http://aws.amazon.com/sns/). With SNS you can push notifications to Apple, Android, Amazon Fire, and Windows devices. Additionally you can send notifications to mobile phones as SMS or as e-mail. You can also send messages to HTTP/S endpoints if you use SNS to send a message through the Amazon Simple Queuing Service. This is a very versatile PNS for pushing messages to cross-platform devices at scale.

The service uses a geographically distributed storage mechanism to guarantee message delivery. You can also use management features to monitor the messages, and if they've been successfully delivered. SNS has been used at massive scale for tens of millions of notifications [46].

At the time of this writing you can send up to one million messages for $1 after the first million on the free tier. (https://aws.amazon.com/sns/pricing/) Well, that is to say you get to publish one million messages and one million message deliveries for $1 to basic mobile devices. You get fewer e-mail and SMS transactions for that. SMS is only 100 for example and $0.75 per 100 after the first free 100. If you had two million devices you needed to deliver to, and you delivered one million messages to each of those two million devices via SMS it would be astronomically expensive. Here's an example of deliver cost to mobile devices using push notifications.

Each message you publish is delivered to two million devices. So in publishing one message you've already gone through your first free million deliveries, and another million deliveries which cost $0.50. If you publish another message it will be two million more deliveries at $1.00. So publishing ten messages is essentially twenty million deliveries. That is about $9.50.

That is if you are publishing identical messages. If you have one message for people on iOS, and a different version for people on Android, that counts as two messages. But you will not be delivering both messages to two million devices, because six hundred thousand Apple devices get one message and 1.4 million Android devices get the other message. So it's still two million deliveries.

Note, this does not include the cost of the outbound data from the SNS service which is extra.

So if you are going to be sending notifications to mobile devices and may need to use SMS or e-mail delivery, or to HTTP/S endpoints you will have to factor in this kind of money for your budget. It's not that expensive for reasonable message publishing. If you are running a massive event like the Olympics where you have to publish to hundreds of millions of devices, or you are a news service that publishes push notifications ten or more times a day to tens of millions of devices, the cost may be a different factor.

When I was researching this topic, I was unable to find a specific service level agreement (SLA) for the Amazon SNS product. While this does not necessarily mean there isn't one, if that is indeed the case, there may not be an SLA provided for the service or it may be a roll up SLA for the AWS public cloud in general. If this kind of reliability guarantee is part of your business process, this may be significant. If you can deal with some periods of outages, then it may be an acceptable risk.

Another Push Notification Service is Microsoft Azure Notification Hubs. They have similar features to the Amazon SNS but they offer more fine grained delivery options based on location and multiple templates for delivery of messages per device type. It also allows you to tailor your target audience with tags to filter messages to specific devices, or to apps and users that meet specific criteria such as "likes this sports team" and "lives in this location." However they do not offer pushing notifications to SMS or e-mail clients directly. This is offered by third party services available as pluggable components in the Microsoft Azure cloud.

The Azure Push Notification Service has been proven at massive scale through Bing News apps for example [47] to deliver "hundreds of millions of notifications" every month to millions of devices. Notification Hubs were also used during the Sochi Olympics in 2014 to deliver about 150 million push notifications to six million devices [48].

Table 5.1 shows a basic comparison chart. Note that this will likely change regularly and so should be considered a point in time view:

TABLE 5.1 Push Notification Service Comparison

Feature	AWS SNS	Azure PNS
Pricing	http://aws.amazon.com/sns/pricing/	http://azure.microsoft.com/pricing/details/notification-hubs/
Send to mobile	Yes	Yes
Send to HTTP	Yes 100K on free tier/$.60 per million	Separate Azure Service
Send to e-mail	Yes 1K on free tier/$2.00 per 100K	Third party service
Send to SMS	Yes 100 on free tier/$.75 per 100	Third Party Service
Send to queue	Yes	Separate Azure Service
Message templates	No	Yes
Management console	Yes	Yes
Route via tags	No	Yes
Scheduled notifications	No	Yes
Specified SLA	No	99.9%
Bulk import	No	Standard tier only
Multi-tenancy	No	Standard tier only
Secure message access	Yes	Yes

The services are fairly comparable. The AWS PNS service has more endpoint targeting built into the service being able to deliver to SMS, e-mail, HTTP, and Queue endpoints. You can accomplish the same thing on Microsoft Azure, but it requires additional Azure or third party services. Microsoft Azure is much more versatile in targeting specific clients via Tags. You can specify up to 3000 tags that message client targets can be filtered by. This includes things like "In Cincinnati", "Likes the Beatles", or Tag Expressions such as (follows_Seahawks || follows_Patriots) && location_Phoenix. This is particularly useful because it allows you to target your push notifications so you can use your limits more efficiently.

Consider the options that these established providers give you when it comes to sheer scale. How much infrastructure would you have to manage to be able to process hundreds of millions of push notifications? What kind of systems and data tracking would you need to keep track of the rules, device and endpoint registration, and message templates? Can you manage the high-availability requirements and failover without losing messages? While you could certainly build something like this yourself, for situations where you want to worry about the functionality of your app, and not the plumbing, using third party Platform as a Service (PaaS) offerings like these can make a lot of sense. In fact it makes a lot of sense in most cases.

As shown in Figure 5.7, you would simply call the SaaS PNS through your SIL. Don't be tempted to call the PNS directly from your Service Layer. If you do later decide to change PNS providers, or create your own, you will have to change all of the services that use the PNS. By calling the PNS through your SIL to route your

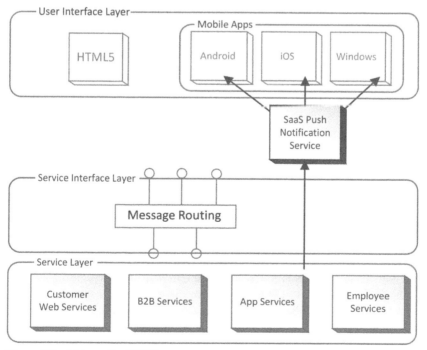

Figure 5.7   Using a SaaS PNS

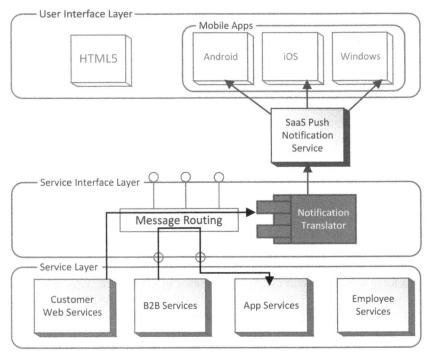

Figure 5.8   SaaS PNS Routed Through TFS

messages to your chosen PaaS PNS, you maintain the abstraction, and you decouple your SL from the PNS. This provides a layer of future proofing and avoids vendor lock-in.

Another aspect to consider when utilizing the SIL is if you are sending messages to other components in your system through the SIL, as you would do in any on premises SOA. There's a good chance that these messages are not the type or format that you would necessarily send to a SaaS PNS. You don't want your services to have to decide if they are talking to an internal first party services so they can send their normal messages, or if they are talking to an external third party service and have to translate their messages.

To accomplish this clarity of purpose, you should keep the TFS component in place as shown in Figure 5.8. This way, you can translate form your native message format into the format specified by your SaaS PNS. The TFS doesn't have to be nearly as complex as it would if you were contacting the various platform-specific notification services yourself. It only has to translate form your native message format, to the format of your chosen SaaS PNS, and forward the message to the service.

You will likely find the TFS a very useful component when communicating with any third party system. It will be used to do the message translation from not only your format to a SaaS PNS, but to any external service that may have its own proprietary message format. Ideally, these occasions will be few and far between if you are using standard protocols and established data formats. Sometimes these are hard to find. For example, while there is an ISO standard for date and time formats, ISO

8601:2004 (http://www.iso.org/iso/home/standards/iso8601.htm), the one for street address format had zero luck getting off the ground. [49] This tends to force other organizations to try to cooperate on a standard, or in the case of postal addresses, organization spring into action like the Universal Postal Union and create documents that they'd like to see used as a standard such as Addressing the world—an address for everyone [50]. Eventually though you will find a message format that is very specific to a particular industry, partner, or user generated service. When you do, you will need to use the TFS to handle it

5.5 MESSAGE TRANSLATION AND ROUTING

One of the primary roles of the SIL is to translate and route messages. In fact, it's the ability to route messages that allows the SIL to decouple service endpoints form the services themselves. This is where you get the versatility and future proofing from. The other main role of the SIL is to translate messages from whatever internal format you may need to integrate with to whatever the apps or external services can understand. These are the kinds of things you want done in the SIL to decouple the translation logic from the services or the clients. The services should focus on doing what they are made for, not translating between data formats just so they can get to the point where they can do their job.

5.5.1 Message Translation

If you are working on Green Field Services for mobile apps that you are deploying, then chances are you know the data formats, and are in a position to use whatever is current and future resistant. So when you are the originator of the services, you pretty much dictate what you want to present to the world. Even third parties who write apps that consume your services will know your outputs and they can code for it.

This changes if you are writing a client for another service, or you are working with brown field or legacy systems. You may not be able to dictate the data formats or message formats the system uses. In these cases you need to be able to convert to and from whatever data or message format that is in use to something more internet, services or app friendly.

For example, if you are working in a hospital and you wanted to use the new HoloLens (http://www.hololens.com) to give doctors access to holographic patient data and ECG charts, you aren't going to be able to send COBOL copybooks and VSAM files directly from the mainframe. You may have to import those file formats into your service, and then your service can send them to the app in a format that makes sense.

You can either do the translation on the way into the service from the Mainframe, or you can do it on the way out to the HoloLens app. If you want to take advantage of any caching of data, you're going to only want to take that translation hit once, and do it on the way from the mainframe to the service and into the cache mechanism.

This has become such a common issue that entire SaaS companies have been built around performing these kinds of operations such as OneSaS (http://www.onesaas.com/). Their core business is providing synchronization connectors that take output from one software or service, and then translate that into a target data format and send it to a separate piece of software or services. When you consider how many different services there are today that you can either access from your mobile apps, or incorporate into your backend services, it's no wonder that people can make an entire business just out of creating the endpoints and message translations.

There are even organizations popping up that just do connector agnostic secure message encryption. You take whatever message you have, and encrypt it then send it via their authenticated encrypted message service. Once the receiver picks up the message, they decrypt it. This can add two additional layers of security, beyond traditional transport encryption, that is not vendor specific to the transport such as Windows Communication Foundation or WebSphere Message Queues.

The first area to consider is translating from one physical format to another. For example, JSON to XML. These kinds of translations can be done through simplistic mechanisms like XSLT, to using built in mechanisms such as JsonConvert in Json.Net, or JSONML.java in JSON on Java. This is the easiest part to do but it requires that the data being translated convert to an identical representation in the target format. If, for example, you have an XML document with one child element that is an array it may get converted to an object in JSON rather than to a JSON array.

You will also have to map from one field or data type to another to fit your view of the data. This is pretty common when converting data types from XML to a relational database system. There are also commercial data integration platforms such as IBM InfoSphere, Microsoft BizTalk, and Oracle Data Integrator. Some are based on an Extract Transform and Load scenarios, others do on the fly translation.

When building your translator service, you need to plan for extensibility as well. You may discover new services to integrate with and won't want to have to build a new translator into the system and re-deploy the translator component. So your translator component needs to be able to read translation mappings at runtime. You should set up a mapping registration that the translator can look up the mappings from based on caller and target. This mapping registration system is filled with the metadata of the message formats from the sender and receiver. The mapping table is a metadata repository. It can either contain all of the information required to translate a message from one service to another, or it can contain information on where to look up the XSLT files, for example.

Some metadata repositories or service bus registries are designed so that each service and its associated message structure are stored as a unit. This causes the translation service to have to look up the metadata for the sender, then look up the metadata for the receiver, and then determine the mapping. Sometimes this mapping is created at runtime based on specialized metadata identifiers, business rules, and rules engines that instructs the translator how to map the fields. This is very cumbersome and takes a considerable amount of effort to develop and maintain. But, it means that all you have to do is add a service to the repository, and all the other services can then integrate with it.

Other systems use mapping documents to map the translations. In these cases, the metadata repository contains lookups for all of the services that each service integrates with. There is a One to Many entry for each service in the repository. So for

TABLE 5.2   Translation Mapping Table

Endpoint 1	Endpoint 2	Mapping Pointer
Service A	Service B	A-B mapping instructions
Service A	Service C	A-C mapping instructions
Service B	Service C	B-C mapping instructions
Service B	Service D	B-D mapping instructions
Service D	Service E	D-E mapping instructions

every service registered, there is an entry for each service it integrates with and a pointer to the mapping information for that particular combination. It can be thought of as a large table as in Table 5.2.

The pointer usually points to something like an XSLT file to perform the mapping, or if you are using a non-XML based system, it will be whatever instructions your system needs to perform the translation. You'll need to use the golden rule to determine which suits your situation the best.

There is a hierarchical structure to the translation. The list below is ordered from the highest level to the lowest level of detail. The types of things you need to map are shown here in Table 5.3.

Notice how they tend to fall into categories defined by the level at which the system is contending with the data. Hohpe et al. [34] refer to them as layers.

This is a preferable term because it makes the translation chaining make more sense when we think of translations that must occur at the different layers of the application stack, even though the actual translation will occur in the SIL of our model. Keep in mind that we are not deserializing a message at this point, we are just translating it from something an external system uses as a message format and data structure, into what our services use as a message format at data structure; custom XML formatted person and address data structure to our JSON person and address data structure. This way, when we are ready to deserialize the data in the Service Layer, it will deserialize correctly

Translation Chains: The chaining ordering happens organically. The first thing the system encounters when it receives the message is the data format, XML for example. So it's likely that it will deal with that first. This could then include making sure the data structures are something we are expecting to see, an array perhaps instead of nested child elements. Once we have that, we can rename fields and data members to what we expect to be able to call in our Service Layer. This has to be last because

TABLE 5.3   Message Translation Layers

Data formats	i.e. JSON <-> XML, ASCII <-> Unicode
Data types	i.e. long to int, string to char array
Data structure and names	i.e. attributes to properties, lists to arrays, a collection of six fields to two separate items with three fields, converting a calculated age to a static age field, Last Name to Family Name, Licens Plate Number to Rego Number, etc)

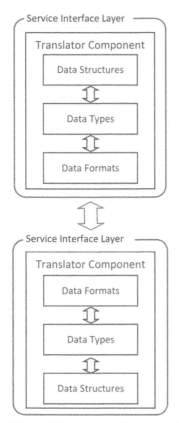

Figure 5.9   Message Translation Chaining

until we have things organized in a manner that suits us, arrays versus nested child elements for example, we can't name the property or node. It may not exist until we address the structure. The chaining essentially looks like Figure 5.9.

The ordered processing of the translations and the separation of each translation level allow you to insert message translations at whatever layer you require, or only use the layers you require. You may only need to do Data Format translation, or perhaps you only need Data Types and Data Structure translation.

The translator component in the SIL needs to be designed so that it can accommodate injecting messages for translation at whichever layer is required. You may even go so far as to put identifiers in the message metadata that say which levels of translation it needs. This may be part of your metadata repository, or built into the messages themselves if both sender and receiver agree to that.

The translator component should contain multiple individual translation elements. They also need to be decoupled as much as any other components do. You will want to use a Pipes and Filters pattern [34] to ensure that messages are sent to only the translation layers they require, and get forwarded to ones that they still need to be processed by. This allows them to be changed, updated, or entirely swapped out without disruption to the actual chaining.

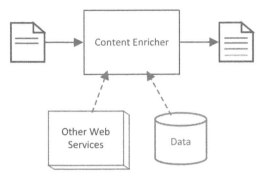

Figure 5.10   Content Enricher Pattern

5.5.1.1 *Content Enricher Pattern*   One of the problems you may encounter is
what happens when the sending service sends a message that doesn't contain all of
the fields you need to process it on your end. There are times when an address may
contain Street, City, and State, but no zip code and you need the zip code for postage
calculations. There is a pattern for this called the Content Enricher (CE) [34] pattern
depicted in Figure 5.10.

When you receive the message that is missing information critical to your system, you have to get it from somewhere. There are three basic ways to do this.

- Compute the information from other fields (calculating age from a birthdate
  field)
- Extrapolate the information from the message, or the environment (timestamps,
  retries, etc)
- Look it up on an external source. (other services, databases, reference data)

The first two are relatively self-explanatory. The last one is the most common
and easiest to deal with. Since the message cannot continue on its way until this
information is added to it, we end up in a situation where the message either needs to
enter a synchronous mode, or we treat the CE as a store and forward system.

Synchronous CE: The CE can receive the message, hold it, look up the information, insert the information and then send the message on its way. This is obviously a synchronous operation. While we want to be as decoupled and asynchronous
as possible in our interlayer, and inter-service communications, within a component
you sometimes need synchronous processes. This also means that the CE becomes a
choke point if the translator component is waiting on it for processing before picking
up the next message from the queue.

Asynchronous CE :In most cases, the CE will operate like other translation
modules and be independent. It's possible that some message won't need to go
through the CE at all. So the Translation Component won't want to wait for the CE
to process a message synchronously just so it can get the next one off the queue and
send it through.

With an Asynchronous CE, the CE itself listens to its own message queue. Messages that have to be enriched are picked up, enriched, and then sent on from the CE

Figure 5.11   Content Filter Pattern

either direct or ideally on a message queue. This frees up the upstream translators, or the Translation Component itself from waiting on the CE to potentially pull information from many difference systems to add to the message.

5.5.1.2 Content Filter
A Content Filter [34] shown in Figure 5.11 is almost the anti-pattern to a Content Enricher. They are designed to strip out information rather than inject other information. Content filters are good for several things. They primarily strip out information not needed by downstream systems. A content filter is one of the ways you can shrink down the contextualized data you send to smaller screen format devices, for example. You can also do this in the service itself and just not return the data from the service to begin with. However a message filter will allow you to have more options in the SIL for message translation, rather than trying to implement all of it in the Service which shouldn't be concerned with message translation.

There isn't a lot of data trimming you should be doing at the Service Layer. If you are restricting outbound data for security, that should be enforced at the Data Layer, long before the Service Layer has to worry about it. The Service Layer will primarily modify outbound data for business rules not stripping it out.

Content filtering also provides simplicity for the receivers which make debugging and processing easier. The less data you have to deal with and continue to pass down the network, the more efficient everything gets. You can also use them to un-nest or flatten hierarchical data. This can be a method of restructuring for simplicity, or converting from relational table structures to more object-like structures. Furthermore, you can use a content filter to flatten and break up complex messages into smaller messages to be sent to multiple systems. This is referred to as a Splitter Pattern [34].

5.5.1.3 Claim Check Pattern
A Claim Check [34] pattern is very useful when you don't want to send really large BLOBs through your messaging system. A Claim Check pattern is essentially just like checking your coat at the theater or your luggage at the airport. As the large message comes in, the BLOS data such as images, video, etc are stripped out and stored in a suitable location. The system then replaces that data in the message with an identifier where the BLOB data can be retrieved from. Figure 5.12 shows how this works logically.

The identifiers aren't always just the URL to the file. If they were then anyone who can either get a copy of the URL, or extrapolate the likely ones and guess a valid one could access whatever data they wanted from whichever messages they wanted. The identifier may even be a collection of data points such as a location, an access key, or some authentication credentials that you digitally sign. If your system permits it, they should also be temporary. The complexity of the protection mechanism is

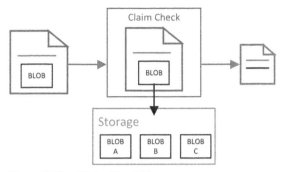

Figure 5.12   Claim Check Pattern

directly proportional to the duration the BLOB has to be managed. If you only have to have that BLOB accessible in storage for a few minutes, then you can probably just generate a URL to the BLOB and name the BLOB with a GUID. Once the duration expires, it doesn't matter if someone gets the link, the BLOB won't be there anymore either having been collected by the recipient or cleaned up by your system.

Many existing systems do this when you sign up to download a publication or a zipped copy of several pictures. The file is stored in an accessible location and a URL is generated. The URL contains the location of the BLOB and its name which is some randomly generated value. Once you download it or after a certain (hopefully short) duration the file is deleted.

If you are keeping sensitive data, and you need to keep it permanently, you will need a more stringent storage process. You should encrypt the data when they are stored and key the encryption to something that is calculated based on information the user can provide. This avoids you having to keep a copy of the encryption key which makes you less of a target and provides a level of comfort to users since they know you don't have the key to decrypt the data.

Additionally, you need to secure the location of the BLOB with authenticated access only. You can do this through user authentication pretty easily, or you can combine that with a shared secret that your systems share with the user similar to standard symmetric encryption, or if you want a higher level of security PKI encryption. You should consult with your IT Security team on how they recommend you to secure the data at rest to meet your business requirements or any regulatory issues you may face.

5.5.2 Message Routing

In most web services exposed on the internet today, you are pretty much talking to the service itself. The service, be that WS-∗ or REST is sitting on a web server and listening on port 80 for requests. You can find it at its URL, and call it whenever you like. This is great, it's easy to access, and it's easy to find. But what if the owner of the service decides to sell it to someone else and retire? What if they decide they want to change their domain from MyApp.MyHoster.net to services.mycompany.net name and they move the URL? Now every client device with code on it that has that URL in the configuration, or worse, hard coded, has to be individually updated with the new address.

One of the things that made SOA easy to adopt was the fact that you could connect your applications to whatever services had been published on the service bus. All you had to do was get the service metadata which described the service endpoints, the available method calls and the data structures to send and receive. This idea was carried through to standard web services through the Web Service Description Language (WSDL). WSDL 2.0 [51], which has been a WC3 recommendation since 2007 and is still awaiting ratification, is in fairly common use and describes the XML format and model for providing information about web services. Some organizations have developed tools for their development environments for discovering and automatically creating instantiable classes to communicate with web services. Some examples are the Apache Axis WSDL2Ws Tool http://axis.apache.org/axis/cpp/arch/WSDL2Ws.html, the Java wsimport tool in the JDK http://www.oracle.com/technetwork/java/javase/downloads/index.html or if you prefer Eclipse there is an Eclipse plugin called ESDL2Java http://sourceforge.net/projects/wsdl2javawizard/ and there's Microsoft's Web Services Discovery Tool (https://msdn.microsoft.com/en-us/library/vstudio/cy2a3ybs (v=vs.100).aspx) as well as tools such as the Web Services Description Language Tool (https://msdn.microsoft.com/en-us/library/vstudio/7h3ystb6(v=vs.100).aspx) to automatically generate WSDL files for services developers created. These tools and technologies give developers a starting place for knowing what format the method calls and payload has to be in to call a particular service, and what they can expect to get back.

In the more recent REST and WebAPI-based world, the discovery of the services and what they require is built into the protocols. The REST services themselves are able to return some data through discovery and hints built into the returns. This is in the form of the HATEOAS structures in the information returned to the client. The HATEOAS approach doesn't describe the entire web service as it is done with WSDL files, but rather each return contains relative links that describe the possible actions the client can take from that point in the system.

For example, you may have an Order Record that was just sent to the client. Given the authorization level of the user, there may be relative links to the Update and Cancel web service operations that are automatically configured to pass in this current order's identifier.

http://www.bhd.com/api/orders/Update/M11328
http://www.bhd.com/api/orders/Cancel/M11328

This structure prevents the client app from having to know too much about the system beforehand. It is a system of self-discovery. Although if you are building out an entire client, it's not always convenient to walk through every possible web service return manually under every condition to discover what the possibilities are and then build a client based on what you've seen.

But in both cases, once the client has the address, you are in trouble if someone moves the service URL. This is where we need to be able to do service routing. We can publish a pretty standard URL to the client, and then publish that endpoint on a service bus that is capable of doing relay work. Then we put the actual service on an endpoint behind that service bus and relay traffic to it. This serves a couple of purposes in addition to allowing us to move things and not update the clients.

We can also route service calls through intermediary components such as input filters and translation layers as I mentioned above. This enables us to perform any message pre-processing and then route the message on to the ultimate service endpoint. This kind of thing is very important for future proofing your services.

Hopefully your app will be a hit and you will enjoy many years of success. With that comes growing, moving, and updating services. You may move from internal to hosted infrastructure and finally to public cloud services. Each environment will have slightly different routing rules and concerns. With the service bus and its routing capabilities, you can accommodate these. For example, what if your company has rules against any inbound communications from the Internet?

Using a service bus hosted outside your organization, you can publish your external service endpoint outside the firewall. Your service can then call outbound to that service bus to establish the relay endpoint that the service bus will route traffic to. You adhere to your company's security policy, and you are still able to offer your awesome services to your client apps. In Figure 5.13, you can see how this would be implemented in the Services Layer.

This will also work in scenarios where you have some services hosted internally, and some in a hoster or public cloud. The same external service endpoints can remain, but their internal target endpoints can be changed to accommodate services being moved to different locations. This combination will accommodate both scenarios where services that are designed for servicing external clients can be routed to directly, as shown by the green line in the Service Layer in Figure 5.14 and internal services that have more sensitive data can still communicate with employees in the field while not exposing the local data center and firewall to attackers.

You may have seen a theme in this book, it's not an all or nothing proposition. You can have services hosted inside and outside your firewall. You may have B2B services that only communicate over a dedicated network like a government or university only network, while your employee services are hosted in-house but reachable from authenticated clients. All while having a public set of services that reside in a public cloud provider to serve your customers and external clients.

To achieve this versatility, you need a good service bus mechanism that can handle external and internal endpoints, with routing capability. You may also need to inject different kinds of translators and filters to massage messages into the proper formats for inbound and outbound systems. These translators need to be modular and only concern themselves with their chosen domain. Updates and replacements are easier and require far less downtime.

Your SIL can take care of cross-cutting concerns like first chance input filtering and security. It can also be used to integrate workflow connections for different filters and aggregators such as the Message Enrichment pattern. This is your mobile app and service post office. All roads go through your SIL including communications from your own services. Even if you host everything internally, the SIL hosted in a cloud or other hoster can act like a firewall. If you put all of your external-facing endpoints in the SIL, and host it in a cloud provider, you can close down all of the inbound ports on your own firewall and greatly increase the security posture of your organization.

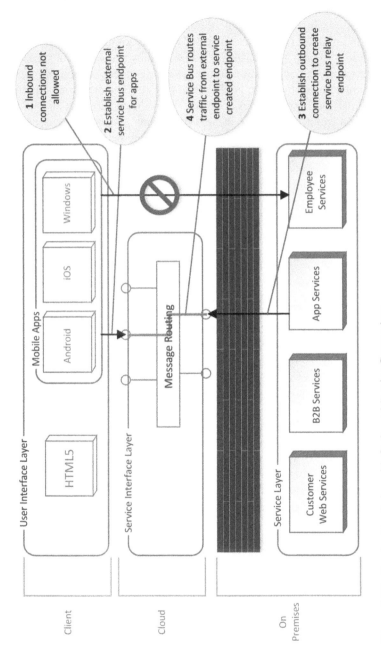

Figure 5.13   Service Bus Relay with Outbound Service Connections

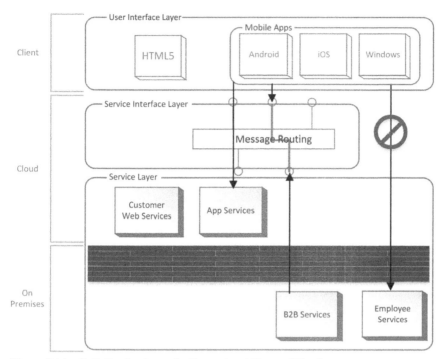

Figure 5.14   Split Service Layer for Internal and External Hosting

5.5.2.1 The Cloud as a DMZ   One of the misconceptions that people still seem to have is that public cloud providers like Amazon AWS and Microsoft Azure are less secure than their own data center. Having visited a lot of data centers, including ones form public cloud providers, I can tell you that in my experience across almost all cases except high-end military and intelligence agencies, the public cloud providers run a tighter ship than you are likely able to do.

People like Amazon and Microsoft spend literally billions of dollars making sure they have the best processes, people, facilities, and security in the industry. They have been researching high availability and security best practices for longer than most of today's internet-based companies have been in business. They have been fighting off attackers successfully for more than 20 years and doing so as high-profile targets. They really do know what they are doing. When you put your services in their data centers, you get all of that best in class security for free as part of hosting with them. This means that you can use their front end infrastructure to bear the brunt of the nasty internet.

Consider this, put all of your externally facing websites, web services, or service bus endpoints in a public cloud provider. If the systems are not data sensitive, put as much of the system in the cloud as you can, SIL, SL, DAL, DL. This will help prevent any latency issues. At this point, if you are able to put everything out there, then you are in great shape. But what about data-sensitive systems or systems that run on legacy hardware like AS-400s or other systems you can't transfer. What if you just want to keep your data where you can touch it?

With some cloud providers, Microsoft Azure in particular, you can create a hybrid cloud scenario that allows you to keep as much or as little of your system stack on premises as you chose, while keeping the public facing endpoints in the cloud. The minimalist approach would be to just publish service endpoints in the cloud and keep everything else on premises. Alternatively if you are only concerned about data at rest, you can keep the data layer on premises and put the SIL, SL, and DAL in the cloud.

In either case, you can connect your back-end servers hosted in the cloud, with your data servers on premises through VPN or dedicated network technologies such as Amazon's Direct Connect, or Microsoft Azure's Express Route. This gives you the ability to have a secure point to point or site to site connection. So the only things that can talk to your network from outside the firewall are the services in your cloud deployment.

In the simplest case, we can implement the service bus relay system as shown in Figure 5.15. Your services can establish the outbound connection like any other PC connecting to the internet so you don't have to allow any inbound connections to your data center at all. Or you could allow them to connect to the service bus over a VPN to ensure that all communications are protected with IPSec.

This same pattern would apply if you hosted the SIL, SL, and DAL in the cloud and allowed communications to the DL held on premises. Figure 5.16 shows this deployment pattern.

Using these deployment patterns you can close down almost all of your firewall ports, and still do business with the outside world. You can take advantage of the cloud providers' best in class infrastructure, security, and bandwidth. Let their experience work for you. At a minimum, the SIL should be hosted in a cloud provider that can offer scale, and best in class security to your applications. The SIL is your service gateway, make sure it's accessible and as future proof as you can make it.

5.5.3 Handling Large Amounts of Data

Sometimes when developing apps, you may need to allow for users or other systems to upload large files such as videos. The SIL itself can be used to pass these large data files but you want to be careful about the channels you use to do it. Large uploads take time even with minimal latency and if you are doing this on your primary service channel it can hold up other operations. So typically if you can do this in an out of band operation you're better off. This usually involves a dedicated file upload capability in your system.

You can handle uploads basically in two ways, something internal to the app such as HTTP-based uploads, and something external to HTTP channels such as FTP. Which one you use will depend on what your platform supports, and what you can do server-side. For example, inline HTTP style uploads will require you to change the server upload limits and server timeouts. If you don't have the ability to do that, this may limit your choice to something external to the app.

A typical way to do this is through an FTP channel. Have the app use FTP to do the upload. This is easier to deal with than HTTP uploads because you don't have to change server configurations to allow large uploads. FTP is made for file transfers.

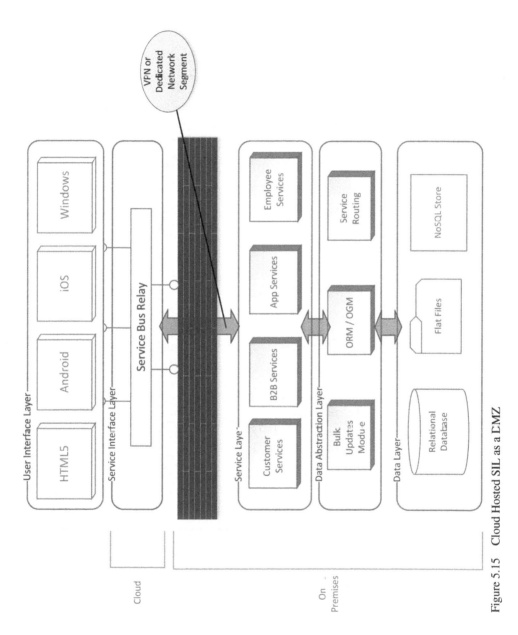

Figure 5.15   Cloud Hosted SIL as a DMZ

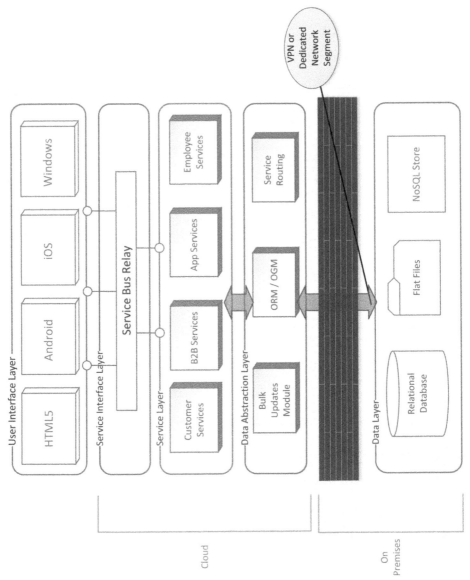

Figure 5.16 Data at Rest on Premises

Most enterprise B2B systems understand FTP as well. But, this does mean something a bit "bolt on" for most mobile apps. Some mobile apps platforms may not deal with this well.

You will also want to consider parallel uploading. Most cloud providers support parallel uploads to their storage platforms. Which provider you chose will determine how the mechanisms work and the details are varied enough to where we won't go into them here, but your provider will have best practice guidance on how to do this. That being said, for anything over 10 MB or so, you'll probably want to implement a parallel upload mechanism.

In both cases, internal and external uploading, you need to consider establishing a direct link to the storage location. As I mentioned above, using a Claim Check pattern, you can establish an out of band upload directly to the storage location, while embedding the location into the message for storage in your system. This will free up your primary SIL to Service channel and allow the upload to happen via a dedicated channel for file uploads. This dedicated channel can even be optimized for upload only and the appropriate security for that one-way path. You'll have a similar path for downloading.

Downloading will probably be done directly from the storage platform through things like Shared Access Signatures, or temporary credentials to download the file directly. This channel can also be optimized for the one-way nature of the download. Large-file handling is one of those cases where the Golden Rule should be applied liberally.

Another way you can upload files is through synchronization mechanisms. This is very similar to how cloud storage providers such as DropBox, OneDrive, or Citrix ShareFile work. User can copy the file to a local folder or server and a background process handles synchronizing the files up to the server, and back down to anyone else that has that share mapped to their local folder. This can often be a good way to share large files and your apps can watch local folders set to synchronize with a server.

This isn't a traditional practice and would only be suitable for line of business kinds of apps. Consumers aren't going to want to have to install a third party piece of software and set up a synchronization folder on their PC just to send you a file.

5.6 WRAPPING UP

The SIL is one of the most critical layers to future proof your apps and services. It is the gateway, traffic cop, translator service, and major access path to your value-added services. It's how you not only receive incoming requests and messages, but how you communicate with the outside world. It is even the path that your service should use to communicate with each other when possible. It contains many of your critical services such as authentication, monitoring, and others as shown in Figure 5.17.

Using this SIL yourself means you know exactly what your customers will experience. You will be dogfooding your Service Layer which means you'll know where it needs improved, or shored up before your customers tell you. This is valuable

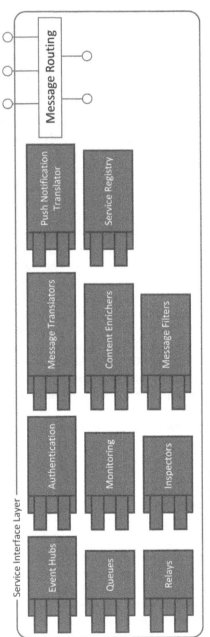

Figure 5.17   Components of the SIL

insight. That being said, there are times when for cost, security, or efficiency that you may want your services to communicate directly between each other. Do what makes the most sense in your situation, but don't overlook the value you get out of using your own products.

The SIL is your first controlled point where you can check messages for security and consistency. It is where you perform your authentication, as well as ensuring that you stop malicious traffic before it gets in the door. You will deploy several components as part of your SIL including your authentication, authorization, and inspection/validation components. These are in addition to the components that will actually handle your messages.

The SIL can also be deployed into a cloud provider without any proprietary information being deployed outside your corporate firewall. It can act as a buffer and security boundary while allowing verified traffic back to your on premises services through a VPN or dedicated connection. Through things like the MS Azure Service Bus Relay you can even block all inbound traffic and have internal services call outbound to register themselves with the SB Relay to pull messages from queues.

Don't overlook the importance of the scalability of this layer. This is where the bulk of your traffic will land. Using monitoring systems to monitor your SIL components can not only tell you how your system is performing, but it can tell you when you need to scale out the backend services to accommodate high loads.

There are many well-established patterns for performing traffic routing and manipulation. Message Translators, Message Enrichers, Message Filters, Message Routers, and Claim Check patterns will all help you set up a highly reliable and highly scalable front door for your services.

By providing this robust abstraction layer, you insulate your clients and external service customers from changes you might make in the Service Layer. They can continue to operate just fine while you add or move services around in the back-end of the system. This ability combined with the ability to change from internally hosted services to external commodity services such as payment gateways also ensures that you keep your agility high and you can always take advantage of the latest and most appropriate technology for your situation.

CHAPTER 6

THE SERVICE LAYER

6.1 THINKING IN NODES

When we talk about services, and websites and supporting our millions of happy mobile app clients, we have to have a frame of reference for our architecture and how we logically discuss the pieces of it. One of the pieces is the deployable part. This deployable part can be a web server, a Representational State Transfer (REST) Service, a database, or some kind of processing system. These components are deployed as a single logical unit of compute or storage. These are most commonly referred to as Nodes. Compute nodes are nodes that perform some software function such as our website, or web services, or some business logic processing. Data nodes are databases, NoSQL data stores, or some kind of file-based storage.

A Mobile Service Node can be the website that hosts your web-based HTML5 client UI, or a REST Web Service, or even the database. It is a logical unit that runs on a virtual machine, either overtly such as a Linux or Windows Server VM or covertly such as a cloud service like Microsoft Azure App Services and Web Apps. These components are the logical unit of scale in cloud-based services. Each of the items in Figure 6.1 is a logical node that can be scaled up or scaled out.

6.1.1 Scale Out and Scale Up

When we talk about scaling, we are talking about increasing the capability of the individual node instances or adding more of them. One of the biggest things to improve with cloud-based services is the ability to scale out applications in near real time. You can scale out your services to handle loads manually, on a schedule, or automatically based on current load. Here we'll discuss scaling out and scaling up our nodes and which to choose to make your services highly scalable.

6.1.2 Scale Out versus Scale Up

We have an interesting conundrum when it comes to terminology now that the cloud allows us to decommission our extra capacity as easily as we can add it. This problem comes from the fact that we have already established Scale Out and Scale Up as terms we use to describe expanding capacity. We can either Scale Up, in which case we add more hardware resources to the individual nodes, or we can Scale Out where we add

Designing Platform Independent Mobile Apps and Services, First Edition. Rocky Heckman.
© 2016 the IEEE Computer Society, Inc. Published 2016 by John Wiley & Sons, Inc.

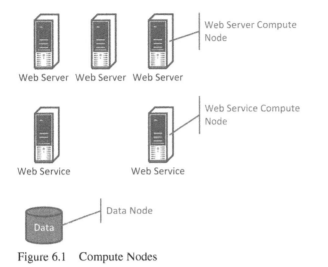

Figure 6.1   Compute Nodes

more nodes to the pool. But what if we want to reduce the resources we have applied to the system? Technically we could say Scale Down and Scale In. Those are perfectly accurate terms, but did you notice how strange that sounded? Part of this is because up until now we never needed terms for those things. It has been a very one-way process. So we need a shift in our thinking. First of all don't look at it like big one-way roads. Look at it more like a graph where we have vertical scaling (Scale Up and Scale Down) and horizontal scaling (Scaling Out and Scaling In/Scaling Back).

It has always been possible to reduce the resources applied to a system. Either redeploying to smaller servers, Scaling Down, or reducing the number of servers handling requests, Scaling In. However, it almost never happens in traditional IT data centers. Once a system is deployed, it's easier to let it sit there with idle machines, than to go through the process of decommissioning or repurposing them. Most in-house data centers had already spent the money on the servers and didn't really see the ongoing costs associated with running them so they didn't bother to scale down, or in. Now that we are working in cloud-based systems, where we pay by the hour for each virtual machine or workload we have running, scaling back operations whenever we can make a lot of sense. Thankfully, it's also very easy to do with cloud-based deployments. So it's probably time to start using those terms Scale Down and Scale In more frequently. Don't worry, the more you use them the more natural it feels.

With that in mind you will see that in cloud-based systems, we don't scale up or down very often. Usually once things are set up, scaling out and in are much more suitable and economical. Horizontal scaling does not require a reboot of an environment or, in the worst case scenario, a redeployment. So we won't see scale up and down much at all. We will see a lot of scale out and scale in. In this book, I will also use the term Scale Back to mean the same thing as scale in. So, what do these things mean anyway?

Scaling Up means that for whatever node we have, be it a web app, service, database, etc. we give that node more CPU, RAM, IOPS or make it faster, stronger, and better in some way. This tends to be a fairly traditional way of handling scaling when we had systems running in data centers where we could just install a bigger machine and deploy the system to it. Scaling Up means that we can deploy the system as-is and it should be able to handle more load. It's easy to do but expensive due to the hardware costs involved. It is also the least desirable for the development teams as vertical scaling also requires longer lead times for hardware upgrades.

This is even more problematic in a fast paced industry where first mover advantage is huge. Development teams need to be able to provision environments, try things, do massive load testing, and decommission environments quickly. If they discover that their system won't meet load requirements unless they add more CPU and RAM, the delay in getting that higher specced server provisioned can mean weeks or months of wasted time while your competitors get further and further ahead.

Scaling up also has a hard limit that is imposed by current technology. Consider when we were dealing with 32 bit operating systems. We had limits on how much RAM a server could utilize because it was only capable of using 32 bits for an address. Then we got memory managers, swappable memory, and software tricks to hide larger memory segments in swap files. Then we went to 64 bit systems which vastly increased our directly addressable memory. But right now, there is only so much RAM, and so many CPU cores you can stick in a server. That is the limit of scale up. What you will find is that scale up works really well and writing software is easy because you don't have to worry about external session state, failover, and load balancers with sticky sessions or server affinity. Then you suddenly run out of head room and you are completely stuck. You can't get any more out of the system until it's re-written for a server farm deployment and is configured for scale out.

Scale out is a great way to get near infinite scalability out of a system. There is a trade-off though. You have to work harder up front when you design and code the system than you do with scale up. You have to plan for nodes being as stateless as possible, session management, externally held session state, failover, and load balancing. If there's anything true about software development it's this; the more effort you put into the architecture and coding up front, the better off you'll be in the future. I've seen several systems now where software vendors have tried to take a monolithic single tenant application, and fork lift it into the cloud by deploying the same application stack that they do for their on premises customers. Then they think that by just giving each customer their own back-end database that they can have a multi-tenant system. But they never designed the session handling, security, and authentication or authorization components in the UI and business layers to cater for that. The system doesn't scale.

There are other benefits to designing systems for scale out. When you design for stateless nodes, you aren't keeping a lot of data in memory, and it forces you to check authentication on each call. This improves the security of the system by keeping the data associated with each call isolated from the next call and the associated user. Plus generally you don't have two callers trying to access the same code at the same time. Each call to your cloud-based services should be single units of execution with their own context. Ideally, each method and function should be thread safe anyway, but you really need to try and get to the point of HTTP statelessness. It will make sure that customer data never collide in memory, there are no race conditions and that one user cannot access another user's data.

As you can see, cloud-based systems, in order to handle massive amounts of mobile client devices, need to scale. Not just scale, but scale quickly, and near infinitely while allowing systems to scale back when required for the best cost efficiency. Vertical scaling approaches can't do this very effectively. They are expensive and have a hard ceiling. Horizontal scaling approaches while being harder to design for due to creating stateless nodes and planning for load balancing and failover, offer that rapid scale out and scale back while having practically infinite scale capacity. The horizontal scaling also minimizes downtime. Scaling up requires reboots and traffic rerouting, scaling out you just add another server to the cluster and put it into the load balancing pool, neither of which disrupts existing traffic.

6.2 PLANNING FOR HORIZONTAL SCALING

6.2.1 Node Sizing

It's all good and well to say that horizontal scaling is the way to go for mobile app services, but how do you do it? What are the considerations? While you could devote a whole book to just scalability, and there are some already, we'll cover the highlights here and leave it to you to sort out the implementation details which will be specific to your scenario and technology and cloud provider dependent.

The key to making compute nodes as scalable as possible is to make them autonomous and stateless. Each node has to be able to operate in isolation. To be true to the nature of HTTP-based communications and web-based systems, each request/response cycle has to stand on its own. If you maintain state information on an individual node, if that node fails the state is lost. Put another way, a stateful node is a single point of failure. It becomes the weak link in the system.

Rule of Thumb: Nodes should be stateless.

Rule of Thumb: Avoid server affinity.

Rule of Thumb: Scaling back is as important as scaling out.

The service tier can be private and can host all of the system's services. Those services that are accessed directly from mobile clients or third parties will have a relay endpoint published in a Service Bus that can route traffic to them.

Remember that smaller VM sizes, to a logical minimum, are better for scale out as they offer a finer grained unit of compute power which will cost less and be utilized better. If you start with 8 core 56 GB of RAM VMs, and you spin up another

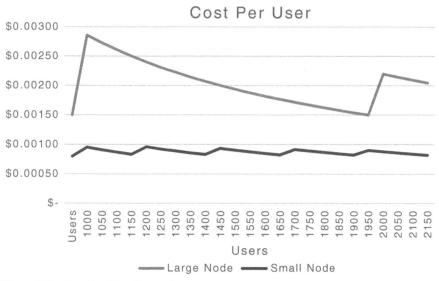

Figure 6.2   Cost Per User Comparison

instance it will be underutilized until the load reaches a point to start to make it sweat. They also cost more in both resources, disk space which means money.

To illustrate the point, if your Large Node can handle 1000 users on a single node that costs you $1.50/hour, when you get to 1001 concurrent users and spin up another node, that other node is only going to have one user on it to start with. Even if you get to 1100 users, that new node is still vastly underutilized and you are paying $3.00/hour. But, if you can get 250 users on a Small Node that costs $0.20/hour, at 1000 users you'll have four nodes running costing about $0.80/hour. Then when you get 1001 you can spin up one more node and only pay $1.00 for all five nodes. At 1251 concurrent users you can spin up another one and only be paying $1.20 etc. Since you can start up 50 or more nodes as easily as you can start up 1, you may as well only start up as much as you will use efficiently. The chart in Figure 6.2 illustrates this. If you anticipate a lot of scale out and scale back in your system, smaller nodes are better.

> NOTE: You need to check the numbers with your cloud provider on how much your nodes will actually cost you, and how many users you can get on a node. These are representative numbers for illustrative purposes. You can't hold me to these numbers.

We all want to be using our nodes as efficiently as possible, and given the numbers above, we can see that smaller nodes tend to be used more efficiently. When you spin up a new node, it is being used more efficiently faster than larger nodes. The chart below is based on the numbers above where we have one Large Node that can handle 1000 users, and Small Nodes that can handle 250 users.

As you can see in Figure 6.3, the smaller nodes are used much more efficiently. Your compute unit to cost ratio is a lot better with these smaller nodes. You will

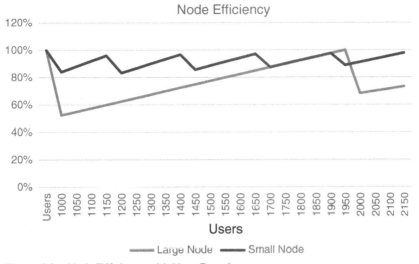

Figure 6.3   Node Efficiency with User Growth

always need at least two nodes active at any one time to ensure failover and redundancy criteria are met. So when you plan node sizing, figure out the best balance for your workloads. Make sure that you aren't over provisioning on the low end of your expected load. If you average 600 users with a peak of 1000, your typical high availability load balanced node count will be around three nodes. In this case, about 200 users per server. This would indicate that your node sizes need to be whatever it takes to run 200 users at about 80–90% of the node capacity. During peak times you'll spin up two more, potentially three more nodes. On average you will run three, and off-peak you can drop down to a single node if your service does not provide a High Availability Service Level Agreement (HA SLA) in off-peak, or two nodes if it does.

With that in mind it is important to put some context around this. Starting and stopping nodes takes time and some cloud providers may charge you a flag fall to start new nodes. If your user load varies wildly in short periods of time, for example, in an hour you fluctuate from 900 users to 1800 and back to 1200, then two larger nodes make more sense to avoid spinning up smaller nodes then shutting them down again only to have to spin them back up.

If your user load has a slower ramp up and fall off time, then the smaller node approach is better. An example of this would be where you have a line of business system that supports users during a typical work day. As users get to work and log in to the system, spin up nodes as they log in. Then at the end of the day when early arrivals start to log out at 3, and later ones at 5:01, scale back.

Obviously you don't want to wait until your servers are smoking to spin up the new nodes. When you set up your automatic scaling, don't set the CPU load to 100%. It's too late at that point. Most autoscaling systems will use an average load over a given period of time. For example, you may set up the autoscaling to spin up more nodes if the average CPU load is greater than 90% for more than 5 minutes. Then scale back if the average load is less than 80% for more than 20 minutes. Scaling back tends to be a lot faster and easier than scaling out. This is usually due to the fact that you can simply take a node out of the load balance rotation, wait for its CPU to finish its current workload, then turn the node off. Booting up a new node, getting its network sorted, and adding it to the rotation tends to take a bit longer so you want to make sure you don't turn off the nodes until you are pretty confident you won't be starting them back up again in 5 minutes. This is especially the case if your cloud provider charges you a flag fall for starting a new node.

6.2.2 Statelessness

For a long time we have had the concept of web farms. These are deployments that are based on scale, high availability, and failover. One of the first problems that was encountered with having multiple web servers was managing session state. Since all cloud-based services are by their nature web based, we need to consider all of the standard web farm best practices when designing our mobile web services.

At its core, the HTTP protocol is designed to be stateless. Each interaction between client and server is one request/response cycle. But in order to emulate thick client applications where a multi-screen form was filled out or a localized set of data such as a shopping cart existed, the state of the user's session had to be maintained over multiple request/response cycles. This state had to be stored somewhere and retrieved to provide contextual information for the current request to give the illusion of a stateful interaction.

In single-server websites, session state is often stored in-memory in the context of the current process. This is easy to set up and very fast to write to and read from. In a single-server deployment if that server went down, the whole site was off-line and the users had bigger issues that if it remembered what page they were on of the current 84 page HR leave request form. To solve the bigger problem we have multi-server deployments that could handle failover if one server went up in smoke. However, this raised the problem of in-memory session state only being available on the server that initially started the session and that server getting all of the future requests from the same user, known as server affinity.

For example, Alice is filling in her justification for personal leave. In this example, let's assume that session affinity is taken care of and all of Alice's requests go to Server A. She might get to page 35 and have to take a lunch break. To this point, Server A has dutifully kept all of her pervious answers in session state in the web processes memory. She goes out for her lunch break. While she is out, Server A decides it wants to go on a lunch break too and crashes. When Alice comes back, she clicks "Next" in her browser window and Server B handles the request because Server A is missing in action, but Server B doesn't know what Server A was keeping in memory in relation to Alice's session. It either returns an error to Alice about her information not being found or comes back with a new empty leave justification form and Alice switches to filling out a Stress Leave application.

When you are planning cloud services for mobile applications, you should not assume that every request will go to the same service node every time. In fact, you should design the system so that subsequent requests do not have to go back to the previous server. If you absolutely must keep some kind of session context, it needs to be kept in an external mechanism that can be accessed by all of the potential servers that could handle the subsequent requests. Usually these take the form of session state servers or a mutually accessible cache mechanism. This is the default approach for cloud-based services.

That being said, you should try to avoid stateful systems wherever possible. Especially in REST-based web services. After all, REST by its nature, and in fact its very name, means the state is contained in the context of the call, not in a session state server or other mechanism. This is where the Hypermedia as the Engine of Application State (HATEOAS) [18] constraint of the REST architecture comes into play. The URL that the request is sent to return a result that contains Relative Links which provide the possible actions a client can take from that point. To provide the necessary decoupling each URL that a request is sent to contains most of the context for REST web services.

All REST interactions are stateless. That is, each request contains all of the information necessary for a connector to understand the request, independent of any requests that may have preceded it. This restriction accomplishes four functions: 1) it removes any need for the connectors to retain application state between requests, thus reducing consumption of physical resources and improving scalability; 2) it allows interactions to be processed in parallel without requiring that the processing mechanism understand the interaction semantics; 3) it allows an intermediary to view and understand a request in isolation, which may be necessary when services are dynamically rearranged; and, 4) it forces all of the information that might factor into the reusability of a cached response to be present in each request. [18]

There are already many good books on REST web services so we won't cover it in depth here. The point is that the more stateful you are, the more you violate the spirit of HTTP, and REST web services which are at the core of mobile app services.

I'm not saying that you shouldn't keep track of what a user is doing over a period of interactions. Websites and web apps have been using things like session state, cookies, and caching to give the impression of stateful systems since HTTP was introduced. But you have to avoid actually creating a genuinely stateful node. You can still fake statefulness by using external session state servers and still do a great job of long running multi-page transactions.

You may receive so many service requests from so many clients that you cannot implement server affinity, or a state server and maintain a decent level of performance. Stateless services are much more scalable and easier to create clients for. When required, state can be maintained on the client and the service calls contextualized to the current state, or on an external server, but avoid stateful services wherever possible.

6.3 DESIGNING SERVICE LAYERS FOR MOBILE COMPUTING

This is the core of a good service implementation and will be a big focus in this book and likely why you picked it up. After all, if you were wanting to build services for mainframe apps you would be reading a different book. So far we have been looking at various techniques and technologies that will allow us to build highly scalable, resilient, efficient, and resource friendly mobile apps and services. But how do we make them as future proof as possible.

We need to focus on designing layers to the system that can abstract away the implementation specifics for the core three layers. While I'm not a big fan of abstraction for abstraction sake, there are good reasons to use it when you aren't sure what the future holds for your system. Isn't there a famous saying by a famous person that there is no computing problem that can't be solved by adding another layer of abstraction?

By putting in these abstraction layers, it gives the system an opportunity to intercept and translate messages (requests/responses, commands/queries) into the format that the layers on either side of it expect. This is what gives us the opportunity to adapt to new protocols, devices, platforms, and technologies. As I mentioned in the beginning, if your service layer is humming along nicely, and then next great mobile platform runs on smoke signals, all you have to do is build a Smoke Signal UI (or let someone else build it) and if they don't understand RESTful web services in order to talk to your systems, you can put a Smoke Signal to REST translator in the Service Interface Layer.

Perhaps you decide you want to ditch relational databases in favor of monkey generated Sanskrit. You just need to put an Object to MonkeyLanguage translator in your Data Access Layer that instead of translating from Objects to SQL, translates from Objects to MonkeyLanguage. Otherwise you'll end up paying some chimp to re-write your Service Layer too.

It's this kind of reactive approach to technology that we are trying to avoid. We want to be sure that when we see the next applicable thing coming over the horizon that we can adapt our service interfaces to work with it. We need this particular abstraction layer to give us that spot to put the code that will do the translation for us. We discuss this in detail in Chapter 5.

6.3.1 Service Componentization

Your service layer itself needs to be designed in such a way that a service is as self-contained as possible. The services you deploy should offer a logically grouped set of functions that fit the app domain that you are providing a service for. You don't want to mix the service purposes. For example, don't deploy a service that combines operations for working with blue hair dye inventory that also handles reseller management. Those are two logically separate domains. While this sounds like common sense, in my experience, organizations tend to deploy one service with multiple operations in a single service implementation. Or they put multiple services into the same service implementation.

Each service needs to be a physically separate implementation to allow for changes in one particular part of the service, for example the products section, to be deployed without having to redeploy all of the other services as well. You don't want to deploy a huge single component that contains all of your service endpoints as pictures below in Figure 6.4.

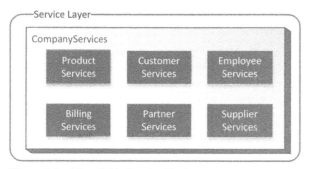

Figure 6.4   Monolithic Service Deployment

If you were to deploy something like this, where all the services endpoints are contained in the single CompanyServices component, if you wanted to change something in the Supplier Services, you have to redeploy all of the other services. This is at best, risky and disruptive and at probably several hours of downtime.

Instead, break out your services into logical domains, and create a component for each of those domains. This is illustrated in Figure 6.5.

This provides you the flexibility to change, update, fix, or remove individual service domains from your service offering without major code surgery and deployment pains. This also makes service to service calls a bit easier to follow and deal with. As I mentioned before, this seems like common sense, but it's scary how often I've seen this in enterprise level services in major industry customer sites.

Often this is simply because they thought all of their company's offerings were the Company X Services. So they deployed them that way. When in actuality, they needed a much more fine grained approach. Categorizing your services and deploying like services together in a single service deployment makes sense. The trick is ensuring that you categorize them at the right level. If you put everything at mycompany.com the level is too high. Instead of deploying one big monolithic service at www.mycompany.com/api, break them down into logical groups www.mycompany.com/api/partnerservices, www.mycompany.com/api/supplierservices, etc. That is at least a bit better than one service. Ideally, you want to break them down to the logical system level. Think at an organizational level, HR, Shipping, Partner, Orders, Field Support, etc.

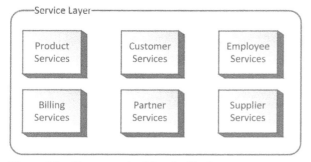

Figure 6.5   Componentized Service Deployment

6.4 IMPLEMENTATION ABSTRACTION

Two of the core principles that provide future proofing for these kinds of systems are abstraction and decoupling. Every time you build knowledge of an external service into an app, other than where to access it and the contract to call it with, you build coupling into your app. This results in an app that will have to be redeployed every time anything inside the service changes. By relying on service contracts, also referred to as service promises, you decouple your app from the services.

One of the ways to facilitate this is to put in an abstraction layer that insulates the app from changes in the underlying services and their physical deployments. At the most basic level, it insulates the app from changes in the location and internal workings of the service. In our case, we implement this abstraction layer between the UI and Service Layer with the Service Interface Layer (SIL). We have a similar layer between the Service Layer and the Data Layer.

6.4.1 Service Interface Abstraction

I should note here that in this architecture and the material that follows, I refer to this as a SIL. It would be proper enough to refer to it as a Service Bus but that's a bit narrow in scope. In most Service Oriented Architecture, there is a Service Bus which is where services are published, and where clients go to find their endpoints. In truth, standard SOA is a predecessor of the design we are implementing, but with some changes to accommodate the highly distributed and multi-platform nature of modern apps.

The reason I continue to refer to this as an SIL, and not simply a Service Bus is because Service Bus has traditionally been a very specific implementation concept. In various SOA implementations to date, a Service Bus was a tangible thing that all services were published to and often accessible only within an organization. An SIL however, can be broken up and distributed across partner organizations, cloud computing tenants, and any combination of on premises and cloud distribution points. It handles both external and internal services calls, as well as intra-service service calls. Most importantly, an SIL offers more in the way of communications and translations.

An SIL can also host your broadcast notification systems, high-speed ingestion systems, and is adaptable to any change in standards, protocols, and service mechanisms. Traditional Service Bus implementations and the lines of thinking that went along with them were not as versatile. So if you want to implement your SIL as a Service Bus, that is completely fine, but consider a Service Bus to be an implementation of one of the components of an SIL, with the SIL being a Superset or bigger picture model than a Service Bus implementation.

At its simplest level, the SIL needs to implement a message router pattern [34] to handle incoming and outgoing message routing. It can be thought of like a traffic cop, a train station, or those big machines in the post office that send letters and packages to the right trucks for delivery. This routing mechanism usually takes the form of an external-facing endpoint that first and third party apps and services connect

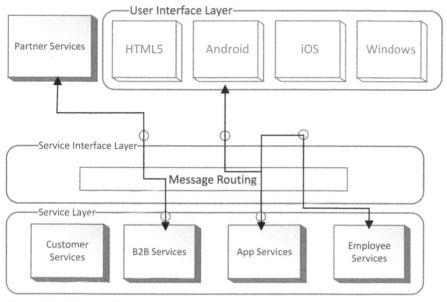

Figure 6.6   Message Routing

to, a message passing structure that identifies and routes the messages to the proper services, and an internal-facing endpoint for publishing push messages, and response messages out to clients.

It's important to remember that when we talk about "external facing" we are talking about it from the point of view of the target services. Services may call other services in the same back-end but should call them through this SIL unless they are purely internal. In these cases, these services are still external, or on the outside, of the SIL from the perspective of the service being called.

Figure 6.6 shows some of the routes that messages will take into and out of our Services Layer (SL) via the SIL. External partner services, and of course the apps themselves will use the public endpoints in the SIL to call our services. Even internal services, such as the App Services, should call the same public, or equivalent privately accessed (secured) end points when accessing other services. In Figure 6.6, the App Services make calls to the Employee Services through the SIL.

This serves two important purposes. First, it makes you use your own service interfaces which give you an appreciation for how easily the services are consumed by external third parties or even other internal parties that have to call those services. Some organizations call this eating your own dog food. Second, it also provides you the same level of abstraction and insulation so that if the Employee Services, for example, were moved from a hosting company to a cloud provider you don't have to change anything in the App Services. You just change the internal endpoint that the SIL routes the calls to for the Employee Services.

It is important to understand this separation of UI and Services. Often when deploying systems that expose service endpoints to the internet, people think of that

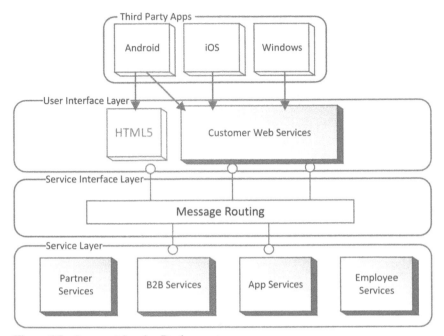

Figure 6.7   Incorrect Service Deployment

as a UI layer. Remember that the UI layer is exactly what it sounds like, it's a User Interface layer. Users don't call web services from their keyboards in the normal course of events. The Service Layer is where you host your web service endpoints. Even though there is a callable endpoint on the service, it does not belong in the UI layer. Figure 6.7 illustrates this mistake.

The thinking that usually goes along with this is that since the Customer Service Web API will be called by third party external apps, that it should be deployed in the UI layer because that's where all of our external endpoints are. But this breaks the model and locks you into a tightly coupled scenario where the external apps become locked to a particular deployment of that Customer Services Web API. You need to keep the services behind the SIL and publish endpoints in your SIL that the external apps can make calls to. Even if the SIL is just performing a traffic routing service, that still gives you the decoupling and app independence you need.

The UI Layer is a logical frame around what could be varied physical websites and apps or even other services. You can put UI websites and apps you create in the logical UI box, and you can even draw third party UI websites and apps in the same box. The important part is that it is for the User Interface elements. Your Services stay behind the SIL. This diagram illustrates the point:

As you can see in Figure 6.8, the clients that make up the UI are all logically in the same UI layer regardless if they are your own clients or third party clients. Anything that is a service, that is anything that does not display data or accept user input should stay behind the SIL.

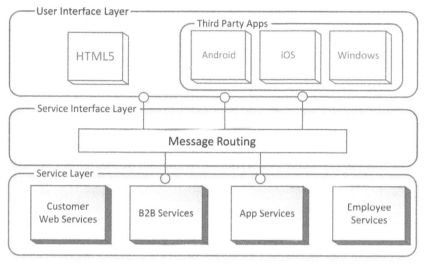

Figure 6.8   Clients Grouped in UI Layer, Services Behind the SIL

6.5 USING CQRS/ES FOR SERVICE IMPLEMENTATION

In order to create a highly efficient and scalable service, we need to be able to optimize the inbound and outbound communications. One way to do that is to have separate channels for inbound and outbound communications. This is a good first step. This prevents services from having to both receive and send messages, and it means that we can dedicate message channels for incoming and outgoing messages. This provides the opportunity to optimize them further. For example, we may need to do authentication on the inbound channel. This makes sense as we only want authorized apps and services to send us messages. However, our services are the only ones that can touch the send point of the outbound channel so authenticating outbound traffic is very easy if it is required at all. Although you probably do want to still do this to ensure someone hasn't hacked your system and is trying to send rogue messages from your servers, but if this is happening to you, you have much bigger problems.

It would be ideal if we could further optimize the messages and handling of inbound and outbound messages. For example, since there isn't much processing on the actual payload of outbound messages, they are usually just results being returned to clients, we can probably limit validation, sanitization, and other processes on outbound communications. On the inbound side, we can streamline it for ingestion and processing of messages.

As it happens, there is an established pattern for doing this. It's called Command Query Responsibility Separation. (CQRS) and we can streamline the message processing with Event Sourcing (ES).

6.5.1 CQRS Overview

Throughout this book we have been looking at technologies and approaches to designing scalable mobile app services. Hopefully, you will need a system that can handle many users across many different devices, with your services operating around the clock. With a little bit of luck, you'll have to contend with efficient read and write operations, data consistency, and high availability. This means you'll need an architecture pattern that provides for all of these things. Command Query Responsibility Separation is here to help. Command-query separation was a term coined by Betrand Meyer circa 1997 and in his book [52]. Then some time later Greg Young defined Command and Query Responsibility Separation as a pattern.

While there have been many articles, books, and even systems designed on this pattern, this is not one of them. However we will go into this pattern in some detail as it is the primary pattern for designing highly scalable mobile app services, especially when we use Event Sourcing as the message passing mechanism. This pattern lends itself very well to cloud-based services due to its extensible, efficient, and resilient design. It is also very well suited to highly distributed systems, or systems where individual segments need to scale independently of each other.

CQRS draws form the Domain Driven Design (DDD) approach to software design. There are many good books on DDD that you can use if you need to get some more information on DDD. CQRS focuses on the domain model at the top end because it describes the business view of things very well. From the app side of the fence, this makes a lot of sense. It also helps us when it comes to the commands we issue to the service because they are common business style commands such as update roster, add new order, update score, etc. When you think about it, your business, what you do, your stock and trade is your domain, so describing the approach to the service design in terms of DDD makes sense.

CQRS at its heart is just the separating of the reading operations of the service with the writing operations of the service. If you had an Order service that looked like this:

OrderService

CreateNewOrder(Order, LineItems)

AddItem(OrderId, LineItem)

RemoveItem(OrderId, LineItem)

CancelOrder(OrderId)

GetOrder(OrderId)

GetOrdersForCustomer(CustomerId)

CQRS would break this up into a Read Side:

GetOrder(OrderId)

GetOrdersForCustomer(CustomerId)

And a Write Side

CreateNewOrder(Order, LineItems)

AddItem(OrderId, LineItem)

RemoveItem(OrderId, LineItem)

CancelOrder(OrderId)

6.5.2 Why CQRS

Since this book is a guide to creating highly scalable and future proof mobile app services, it's natural to wonder why this pattern is suggested as a core tenant of the approach. There are a few reasons that I recommend CQRS as the pattern to use in your service layers.

- It focuses on taking your app ideas and business domain and creating it in software
- It provides independently scalable read and write systems which are ideal for very high load systems.
- It is a very flexible system with components that can be independently versioned, changed, swapped out, and scaled without affecting the overall operation of the system
- Separating the read and write responsibilities reduces the complexity of the code and data management components you have to write and reduces the impact of changes in business logic to fewer components (usually on the write side)
- It improves testability and debugging through the isolation of the read and write operations making stubbing out these operations for testing easy
- Being able to add or change features is easier because it will only be applied to one side or the other. New screens or reports won't affect the write side and new user capabilities won't affect the read side

When you start thinking of things in terms of a Read service and a Write service, it opens up a lot of possibilities for making a very efficient system. One of the big advantages is that the data model on each side can be optimized for its purpose. The Read Side data model is entirely read only. There are no locks required on anything. The write model will process commands sequentially so there are very few if any locks required on the Write Side. Another advantage is that the data models on each side do not have to be the same structure. While this sounds wrong at first, let's go through the idea.

6.5.3 Being Able to Separate Data Models

It is important to keep in mind when discussing data models that the physical data storage mechanism is whatever you want to use as long as it works. While we talk about relational databases, NoSQL Stores and flat files, it doesn't matter. This is why we have a Data Abstraction Layer over the top of the Data Layer. If you want to persist incoming commands in an NoSQL data store, and then have the results of the commands persisted to a relational database for the read side, that's fine. The actual implementation details are up to you. Do what makes the most sense in your situation.

In many systems you have a data model that normally consists of a relational set of tables, or some kind of file persistence store that tends to be object based rather than relational. In almost every large service-based system, there is a relational database hiding in the back somewhere. Relational databases by their nature are normalized.

That is to say the data are broken up so that it is not contextual but it is very easy to avoid duplication. Instead of having one large Customer Order table, you have an order table, a line item table, a customer table, an address table, and several other tables that hold reference data that order line items draw their constrained data from such as colors available, shipping partners, fabrics, models available, etc.

Taken this way, the data don't make sense to a user. The data have to be combined into some contextually sensible structure to make sense to the user. This is why we have stored procedures and views in databases. If the user wants to see an order, the system pulls all the information from the relevant tables that relate to a particular order number. Once it has all of that information, it sends it to the app where it is usually turned into some kind of in-memory data set or an object. The app UI then displays the various parts of the object on the screen in a contextually useful manner. Even though we have all the data we need in the database, we still transform it into some combined form so the user can make sense out of it. This process happens every single time. So why not have an optimized read side for these operations?

The Read Side of the CQRS system pulls data from stored procedures views and ready to present sources. These data are also guaranteed to be read only as it comes out of the system so you don't have to put a lock on anything. Don't worry, we'll get to updating the data and not having to worry about Read Only when we talk about the Write Side. This means you can not only cache data efficiently, but you can also precompile all of your queries, and even create denormalized tables specifically designed for reporting and read operations. This means that the service itself has to spend less time to actually assemble the normalized data into something the user can understand and you have to write less code.

On the other side of the pattern we have the Write Side. As you can probably guess it is optimized for write operations. Since we know that the only thing we will be doing on the data store on this side of the pattern is writing to it, most of the work will be insert, update, or delete operations if we are using a CRUD style data-centric system. The data side service can be designed to understand and works very well with normalized data structures. With the service-centric model, we can queue up the commands sent from the user and process them sequentially to avoid any data locks. We can also persist these commands and have an instant audit trail.

The app itself can be modeled so that it tracks what the user is intending to do, and then sends messages to perform that action on the service. UpdateOrder, RemoveItem, AddItem, CancelOrder, etc. are all things that demonstrate user intent. In fact when designing your method names or services, put "The user wants to" in front of it (in your head please). If you do that, before you know it you'll have a Task-Based UI and you'll be ready for Event Sourcing which is where we're headed with this.

This separation of Read and Write sides also provides you an excellent security improvement point. Since all of the write operations are happening on a very specific service boundary, you can secure write operations at the very top. Read operations will be the same. If you have an OrderReadService and an OrderWriteService with their appropriate methods, you can apply your security at the service level which makes it easier to manage than applying it at each method. Although, you should probably do that too for a good belt and braces approach (or belt and suspenders for those of you in North America).

A very important concept to understand with the separation of the read and write side data is that you have a scenario where you will have Eventual Consistency. This means that there may be a delay between the write, and when that data are available for read. In most cases, especially large internet-based distributed systems where there is an expectation of some delay, this is acceptable. You need to apply the golden rule here to see if your business rules are ok with Eventual Consistency. If your system needs near real time visibility of writes, then you may need to combine the read and write side data store into a single entity such as a relational database. This will allow the write to be visible on the next read that is not from cached data or pre-built contexts.

6.5.4 Aggregates and Bounded Contexts

When working with CQRS, there are a couple of terms that are used to define the domain we are referring to and the things in that domain we are working with. One of the terms used in DDD is Bounded Contexts. This is pretty much what it sounds like, it is the group of data, information, and operations that are specific to a subsection of the domain we are working in. You can think of it as a sub-domain for the domain model. You may have a few of these, or you may have lots of them. It depends on your business domain. The best thing to do is apply common sense with a healthy application of the golden rule. What is a logical boundary that you would define a service on?

In the case, for example, in our escalator company, they may have a domain for installations, one for maintenance, and one for research into new fields such as thrill rides. Another example for an online retailer is OrderTaking, StockManagement, AdsAndPromotions, Warranty, etc. I'm sure you probably already have a good idea of what these bounded contexts will be based on how you've drawn up your ideas. Consider your use cases/user stories as a good source to discover them. Or you can think of it in terms of your system boundaries as seen from the outside such as the HR System, KPI system, Training System, etc. each of which may have their own subdomains. However, as mentioned in Eric Evans' book [53] you should prefer larger and fewer bounded contexts.

Within those bounded contexts, we have concepts of Aggregates and Aggregate roots. In order to avoid having to lock an entire database to perform an update we need to know what an update is likely to affect and what it won't. Defining that reach is the job of the aggregate. Think of it as a way to define a boundary that lets you ensure data consistency. An aggregate is a group of entities, value objects, and other integrated data that need to be taken together and changed as a whole to ensure transactional consistency. For example, if you have a Customer object, it may be made up of an Id, one or more addresses, phone numbers, history of correspondence, next of kin, etc. That customer may also have many orders associated with it. These may be grouped together in a CustomerRecord. The Customer is an aggregate, the Orders are aggregates, and the CustomerRecord is an aggregate made up of two other aggregates. An Aggregate Root is the mechanism that is used to access all of the objects within it. You need to have access to a customer to be able to access the Customer data within it. The aggregate root is the thing that you create to get access to the data inside it. You don't create an Address, try to get at the Orders for the customer that belongs to that address. While this is possible in a database query it is not how the architecture and design of a business context work. If you update an aggregate, that update will potentially affect all of the objects, data, or aggregates within the aggregate you update. So an aggregate defines a boundary around the things that could be affected by changes to one of the items.

6.5.5 The Read and Write Sides

As mentioned, applying the CQRS pattern is a matter of separating your read and write operations. It's that simple. Figure 6.9 illustrates that.

The red arrows represent the one way inbound commands on the write side, the darker arrows represent the one way outbound read-only data on the read side. The gray arrow represents the write side informing the read side of the change based on the commands that have come in.

It is important to note that this is a logical separation. In the diagram the two data stores are separate things. This may be the best solution, or a single data store may be better. It depends on the circumstances. If they are separate, you have to find a way to communicate the changes to the read side. In a data-centric design this may be as simple as the read side, creating its views from the same data that the write side is writing to. In an Event Sourcing model you may want to store the events as they come in and publish updates to the read side data store by way of events or messaging.

Update events can be raised from the service command handlers themselves, or directly from the Data Layer. In each case, you are creating a transactional boundary to ensure consistency. Once the write succeeds, the read side must have access to it. So if you issue an insert command to a relational database that is shared between the read and write sides, the transaction is limited to the successful insert. If you are using a command-based system where the commands are stored as they come in, once the command is stored the read side needs to persist the change to its data store. So on successful execution of the change command, the read side notification needs to be sent as well to consider both sides of the transaction complete. To go a step further, the read side can return an acknowledgement to the write side if guaranteed messaging is implemented. We'll cover this in more detail later.

6.5.6 CQRS Communications

One of the primary aspects of CQRS is Commands. It's even in the title of the pattern so it must be important. When you think about the app you want to create, you think in terms of the things a user will do. After all, you're not going to be creating an app and its associated services if there is no user. Once you have that information, you sort out what data you will collect and utilize, and how you will utilize it. Alternatively you may start with a large collection of data, and create a service that gives people or other systems access to that data. It may even be system to system and not have users. But you still need to know what that other system is going to want to do. Unless

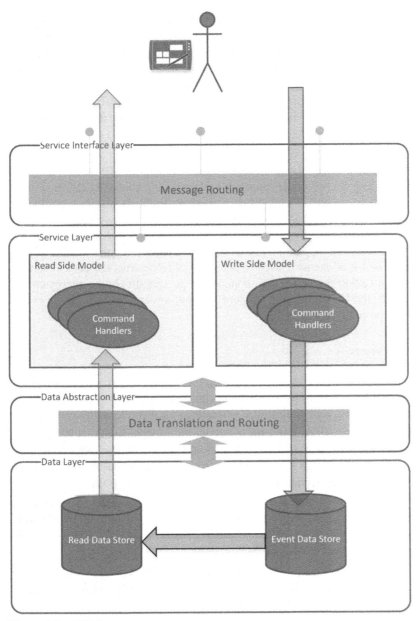

Figure 6.9   CQRS

your system is entirely a static one-way publication of data, it will need to accept and act on instructions to know what to do. So in all of these cases, we have to accept commands. Once the service has accepted the various commands and performed its work it needs to tell other interested parties, users, or machines, and even the read side that something has happened. This is usually done with an Event.

Events are notifications that are either sent directly to an interested party through a message queue, or are broadcast to multiple interested parties through some kind of broadcast notification or a publish/subscribe model. These messages are how the service communicates. When the read and write data stores are separate, you can have the write side system write the message to the write side data store, and to a notification channel to get the read side to update itself. You may also choose to have the write side send messages via an event to notify the read side to update its data this is called Event Sourcing.

6.5.6.1 *Event Sourcing*

Event Sourcing is the ES in CQRS/ES. ES is an important concept to have a good understanding of when talking about CQRS/ES systems and separated read and write sides. In this context an ES is not just an event like an exception or a notification that something happened somewhere to some object. In this context, the ES is actually the record of the actions performed by the user, and stored as the transactions applied to the system. ES is part of the current state of your application. The list of events is used to construct or reconstruct the current state of the system from either the beginning, or since the last snapshot of the aggregate was taken. As such, the ES is the history record for your system and as such the events require a few properties.

6.5.6.2 *ES Events*

- Are immutable – they have happened and are stored as a record of actions in the system so they must maintain their original form. This is important because they can become your audit trail if you make sure this rule is enforced.

- Represent one-way communications. Aside from implementation level details such as guaranteed messaging on message queue technologies that send acknowledgements that a message was received, the sender to receiver communications is one way, there is no response to the event itself.

- Represent the user intent, and contain all information necessary to execute that intent. "Alice cancelled order 44349," an event should not be "UPDATE Orders SET OrderState = "Canceled" WHERE OrderId = 44349" as this does not describe business intent to non SQL experts.

- Can be raised by aggregates, the service or the data store itself. Ideally in a service-centric system the aggregate will raise the event and include its aggregate ID to identify the source of the event.

Now that you have context for what an ES is, we can discuss how it is used as your systems' state. When you think of the current state of a system it got to this point by the data starting at some value. Over a series of changes to the system it got to its current value. The record of those changes is the list of event sources that have come into the system since the start point. Table 6.1 demonstrates this.

The Blue Hair Dye aggregate will start off at a quantity of 45. As each message comes in, it applies the change to the quantity. If a snapshot is not taken, the next time the system starts, it will read the last persisted final value, zero if there were

TABLE 6.1 Event Sources

Original Value:	Qty Blue Hair Dye Available: 45		
		Sell 2 Blue Hair Dye	Append
		Sell 1 Blue Hair Dye	Append
		Stock 15 Blue Hair Dye	Append
		Sell 15 Blue Hair Dye	Append
Current Value:	Qty Blue Hair Dye Available: 42		

no snapshots or 43 in this case from the last snapshot. Then it will replay all of the messages since the last snapshot to recalculate the current value.

If an event needs to be changed, it cannot be modified in place. A reversal command that undoes the initial change must be sent to the system. While this sounds cumbersome, remember that it keeps your audit trail intact. It is also a fairly common practice even in other fields like journal accounting. If a credit was accidentally entered into the system, a debit must be entered to reverse the journal entry. This practice actually gives us very good performance improvements as well. Because events are immutable, they cannot be changed or deleted. So you never have to lock the event source data store. All events are only inserts. This means you don't have to worry about locks on the write side or the read side. That is not to say that if you are updating a data store, you won't have conflicts, but we'll address that shortly.

Another benefit that is worth noting is that because you have a list of all of the transactions performed on the system, and they are designed to show user intent, you have a gold mine of Business Intelligence you can get from this information. Rather than having to add separate BI tracking information into the system so that it records the users' actions alongside the data, the events themselves become that information for both sources.

Events are also very well understood mechanisms that can be copied to other systems. This makes system to system integration quite easy. With the proper DAL in place, the information in the events can be translated to target system data formats, or even changed entirely for the target system. Note, I am not saying you change the events in your event data source, but you send a copy of the event and translate it from say JSON to binary serialized data or a SOAP/XML package. The events themselves can be converted for storage in any persistent store.

When you must have the event persisted to the event store, and published to the read side in separate operations, you should use a two-phase commit transaction that binds the message write and message broadcast operations together. You can't have a situation where the data are published to the read side but the write side fails to write the message into the data store or vice versa. This bit of advice needs to be tempered against your need for high speed. Two Phase commits impose a performance hit on the system. In fact it can limit the top end of your performance. So make sure that if you need to do this, that you don't cripple yourself performance wise.

Considering those issues with the write side there is a trade-off triangle as seen in Figure 6.10. For your write side you can essentially focus on two of the points, and accept the third one. You will be able to get high performance and low sync lag with published messages. You can get high performance and reliability with asynchronous

Figure 6.10   Tradeoff Triangle

database sync. You can get reliability and low sync lag if you use the two-phase commit version of message publishing. Depending on your services' needs you will have to find a balance point on the tradeoff triangle.

Regardless of the approach you chose, you have to accept the idea of eventual consistency. As I mentioned before, in most internet-based systems this is a modern expectation. The efficiency and code simplification you get out of it makes it a pretty good tradeoff.

6.5.6.3 Commands and Queries

In most current distributed applications, data are often sent to clients in some form of data transfer object (DTO) like a DataSet. The client performs all of their operations on the DataSet then that DataSet is sent back up to the service to be synched with the back-end database. This works pretty well and provides a very good offline experience. In fact this is probably what you were thinking of doing when you were planning your Line of Business app. You don't have to toss out that line of thinking when you use CQRS/ES. You can still transfer DTOs from the read side to the client, but when you send the changes back to the service, you will send commands instead of the entire DTO.

Commands in CQRS are how we tell the service what we want to do. They are received by the service and processed by a command handler. I gave some examples of commands earlier when I talked about Event Sourcing. "Alice cancelled order 44349" was the event. Cancel order 44349 is the command. There are a few things you need to avoid when dealing with commands. First, commands need to stay within the bounded context. If you are talking about the HR Leave System, it doesn't make sense for that system to send HR leave commands to the Car Leasing System.

This works as far as the service side goes. The mobile app isn't part of one particular bounded context and may in fact need to send commands to several different services, all of which belong to different bounded contexts. The app itself may have an HR Section, a Work Reporting Section, and a KPI section. All of these are examples of different bounded contexts that are accessed by the single app. So for this reason it's easier to keep the UI above the bounded context and design it as sending commands into the bounded context.

You may also have other services, or orchestration systems that send commands to various command handlers, or from one bounded context to another, but

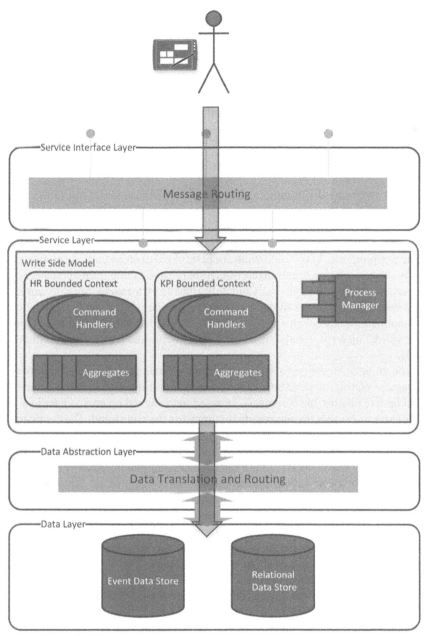

Figure 6.11   CQRS Write Side

they should be specific for the purpose and act as more of a Service Bus or workflow routing system and are not considered part of the bounded contexts they communicate with. These are often referred to as Process Managers. Figure 6.11 shows how these pieces would be logically deployed.

When you design your command handlers, you're going to want to keep them logically separated so that they perform similar functions. The natural way to do this is to have a command hander for each aggregate.

As you might expect a command handler does exactly what it says it does, handle the incoming commands related to the aggregate. You will have a handler for each type of command that comes in. Depending on your style, you can either do this through separate methods that are named for the commands they process or have an overloaded "ProcessCommand" method that takes different command types at its parameter.

All command handlers operate pretty much the same way with the same responsibilities:

It must validate all commands. This is crucial because this is one of the first chances your app has to perform contextual deep validation of the command and its parameters. Ensure that you are doing the best job of validation that you can. In some cases you may even build a dedicated parser to check for things like SQL Injection, Cross-Site Scripting, etc. You should think about this kind of cross-cutting component in the early stages of design. Centralized validation is a good thing from a consistency perspective and it lets you apply the validation rules from a known central location instead of spreading out copies of those rules to each system. This makes them much easier to maintain. It really helps in avoiding security vulnerabilities that arise from fixing the validation error in only nine out of ten systems and leaving that tenth one vulnerable.

If you are lean on the performance side, and you can afford an extra shim you might consider putting in a command pre-processor that does all of the parameter validation up front before the command is passed to the actual command handlers.

It locates or creates the aggregate instance for the command. The command handler needs to access the aggregate via the aggregate root to process the commands. To do this, it either needs to activate the instance that is already operational, or if none exists it must create one. It is important to remember that if you are using Event Sourcing, this is the point where you would re-hydrate the aggregate. You would get its state from the last snapshot, then replay all of the events since that point to rebuild the current state of the aggregate. If required you would version check the data here as well to ensure the commands are not operating on expired or changed data. You can do this through checking the timestamp of the data or the last event processed, and compare that to the command about to be processed.

Process the commands using the Aggregate Root methods. Once the process manager has an instance of the aggregate it calls the methods on the aggregate root to process the commands. This is the core the process managers job.

Save the modified aggregate. Once the command has been processed the command handler then saves the aggregate in its new state. This is most likely going to be in some permanent storage accessed through the DAL. This is the point you publish the events for the read side, or any other interested parties. Since the data persistence can be abstracted from the Service Layer, your DAL will decide how to persist the aggregate to storage. If you are using Event Sourcing, the DAL will pass the new event to the Event Store for the aggregate. If you are using a database or ORM model then the DAL will save the aggregate information in the underlying database. In either case, the DAL simply accepts the "SaveAggregate" call and then persists the change in whatever mechanism you have chosen in the data layer.

Depending on how you want to structure your system, this is also where you would snapshot the aggregate on a regular basis. For example, you may want to snapshot it every hundred changes, or on a time basis. Depending on the complexity of the changes in each command, if you start seeing performance problems because the system is rebuilding the aggregates and processing the messages is taking too long, you might want to shorten your snapshot duration. The more frequently you take snapshots, the less the system has to process to rebuild the aggregate state.

6.5.6.4 Concurrency
One of the issues you may face with this pattern is concurrency problems. Like any system, the fewer the users the less likely you are to have issues with concurrency and keeping consistent data. In small systems you can have one of each aggregate behind a single queue and directly control the changes by processing one message at a time. While this won't entirely eliminate consistency problems, it helps. If someone has read the state of an aggregate, and right after that someone else sends in some commands that changes the state of the aggregate, when the first user sends in their changes, you have a conflict because they are not working on the most recent version of the data. We talked about this before with Alice Bob and Charlie. The solution there was to use data versioning. This keeps data consistent on the service side, but will result in some users having to re-try their operations. While this is not ideal from a user perspective, it's better than having inconsistent data in the system.

If you have high-load systems, you will need multiple instances of the aggregate to handle the incoming commands. This will result in some parallel processing and will inevitably produce conflicts. In these situations you have to implement a versioning scheme to keep a level of sanity on the data.

You also need to consider that in a cloud-based system when you are duplicating data across geographies, or even mirrored systems in the same region, you will have period of inconsistency. This is always going to be the case while the data are synchronized between sites. This is assuming you have a single central data store that all the transactions occur in. In a globally distributed system, this will be much more complex.

If you are serving customers around the world, you probably don't want them all operating on a single data store located in Australia. You will want geographically local, and likely localized versions of the data close to the locations the customers are in. This makes sense and if all of the customers are localized, then you don't have a problem with cross-geography data consistency.

However, you may have a system that allows everyone from around the world to interact with not only your system but with other users around the world. You may allow users from North America to enter data in systems located in France. You may even have situations where multiple users can enter information into multiple records in multiple regions such as global check-in sites, travel reporting sites, world sporting events, and other globally distributed systems.

In some cases, simplicity is best. Even Facebook and Twitter use a centralized data store. In the case of Facebook it's a large Hadoop cluster with Hive sitting on top of it [54]. They've had to highly optimize the system to handle globally distributed loads, but it works very well.

So don't' get too fancy with distributing your data stores and coming up with a complex mirroring system that deals with conflicts and data sync issues. It might be best to use content delivery networks and optimize your systems to handle data very quickly. You may be served best with a high-speed file system storing incoming messages and running Hadoop over them to manage the data. Don't' blend things together if you don't have to, and don't create problems for yourself.

> *Rule of Thumb: Favor simplicity in all things.*

Don't create multiple data stores if you don't have to. If your customers will only be operating in the data store in their geographic region, then put one there. But if you have to combine many entries from many places around the world into a central intelligible data system, then get the data to that system. Liberal applications of Occam's razor will do you a lot of good here.

6.6   SIDE BY SIDE MULTI-VERSIONING

One of the things that will inevitably happen is that you'll get a mobile app deployed and it will happily talk to its associated service. Then, someone will want to change something. This is as sure as the sun rising. What happens when you want to add new features, or change something? What if you deploy two different versions of you app, a Trial and a Pro version? You might need slightly different services, or perhaps they can call the same service, with slightly different results. What if customers build their own systems that integrate tightly with yours and when you are ready to move on to bigger things, they aren't?

At some point you will likely have to run different versions of the same service in parallel. This has been a problem in computing since DLL Hell. Some of you may be fortunate enough not to have encountered this, but for those of you too young to remember there's Wikipedia: http://en.wikipedia.org/wiki/DLL_Hell

Howard Dierking has a great post on versioning REST services [55] that I tend to agree with. Your versioning scheme will depend on what it is you are changing. There are three types of changes you can make, additive changes, format breaking changes, and complete changes in which the services essentially do something different, or return different information than the previous version.

In the additive change, you are pretty much ok as long as the clients are designed to ignore what they don't understand. If there is a new operation to call they won't be calling it because they don't know it exists. If they get extra data they will ignore it and just use the data they were expecting. This does kind of violate the rule of "only send the absolute minimum data necessary for security and bandwidth purposes" but we will also be servicing new clients that will require that information.

In the format breaking change scenario, you need to version the data format and use content negotiation and the accepts header to ensure that the client is getting the data in a format it can handle. Otherwise the clients will simply start malfunctioning and your ratings will magically reflect that you had a lot more users than you thought you did. People love giving bad ratings to things, but they don't ever seem as forthcoming with the good ratings.

In the last case, where you've effectively created a new service at the old address, you need to actually version the URL. http://www.bhdcompany.com/api/ order/V2/blue/cornflower/42 Versioning the service this way lets the existing clients continue to call the original version and expect the original purpose and results from the service interaction. New clients can use the V2 version with the new context.

6.7 SERVICE AGILITY

6.8 CONSUMER, BUSINESS, AND PARTNER SERVICES

With the common integration of Business to Business (B2B) and partner organizations now, it is important to understand that you need to plan early for integration with external parties. Even if you don't have any or don't integrate with external parties now, you still need to lay the groundwork. This has been the crux of many problems with B2B integration to date.

In the past we've had all these systems that organizations relied on. They worked just fine, as long as you were in the same building as the systems you needed to use. Then we decided that these systems needed to talk to systems in other organizations. So we joyfully connected the networks together through modems, and eventually direct connections from our basement to their basement. Then we discovered the Internet. Once we found that, everyone was scrambling to see how many of their internal corporate systems and customer service systems they could bolt a web interface on to and publish to the World Wide Web!

All of a sudden, we had people accessing services that were never designed to be accessed by people who didn't know the nuances and quirks of the system. We had to handle traffic that on a scale we hadn't seen before, and we had to deal with bad guys, lots of bad guys trying to break our stuff. In short, chaos ensued. If you don't want to repeat that cycle, then you need to plan early for integration with other services.

You might want to expose your services directly to consumers. This is the case with things like Facebook and Twitter. Anyone can write an app that accesses the service APIs for these social networks. These are very consumer-focused services or were so in their inception. But like any good entrepreneur, where the people are, advertising will soon follow. So these two social media giants added advertising and demographic APIs to their list of services. Wise move, especially since they were going public with their IPOs.

This brings in the business to business (B2B) or business to customer (B2C) services. You may collect information on your customers that, when properly de-personalized, may be very useful to a demographics agency, or advertising agency. They will want raw data and your services will be called by their services and systems. Not necessarily a person using a web browser. These third parties may want data in

XML, or JSON, or OData, or even binary. Somehow, you have to be able to adapt to that.

Keeping the previous discussion about the SIL in mind, we need to plan our services in a way that they can be independently deployed, monitored, and updated. We also need to plan for restricting access to a subset of our services to a subset of our customers and partners. This will lead into a discussion on authentication and authorization. After all, you can't decide if you need to allow a caller to use a service if you don't know who they are, or what they are allowed to do.

6.9 PORTABLE AND MODULAR SERVICE ARCHITECTURES

When building services or any computer-based system you have to pick a technology. I mentioned this early on, but it bears repeating. At some stage, you will pick your first hosting platform. It may be in-house, it may be with a hoster or with a cloud provider. You need to do this to get going and have something to deploy on. Over time though your requirements might change. Usually it's the scale capacity and cost that need changing first.

This means that you may have to move your services from your initial deployment platform to a new one. You may move from a hoster to a public cloud provider or from one provider to another. You may even move from a public cloud to a private cloud in-house. In any event, your services need to be portable.

Portability is one of those things that requires a measured approach to. If you design for complete portability, your service will be average at everyone and not spectacular at anything, other than being portable. The reality is you probably won't move service around very much at all. Sure, you may have to do it, but it's pretty unlikely.

I've seen several startups that decided they wanted to design their system for the multi-platform world. I watched them grind to a halt in analysis paralysis because they couldn't find a combination of code, libraries, hosters, and developers that didn't involve some form of technology lock-in. We need to keep a dose of reality involved in our decision making.

With that in mind, what do portable services look like? There are a few things that they need to "be"

- Self-contained – If a service spends all of its time calling out to other components or relying on specific data structure knowledge in the database, it's going to break when you move it or change something. Think of it this way, you should be able to deploy the service pretty much by itself, and it should be able to do everything its domain requires of it. Internal libraries that are built in or referenced are fine because they are deployed as a unit. It's part of the service. But if you have to deploy an entire different service or stack just to get this one service to work, you need to reconsider the architecture.

- Data implementation agnostic – Your service should not know anything about the physical implementation of the data storage mechanisms. If your service is building dynamic queries and calling specific table and column names, it's broken. That is a job for the DAL, not the service itself. If you build structure-specific information into your service, you can't make changes to the data layer without testing absolutely everything above it and rebuilding and redeploying your service. Your service is inexorably tied to the physical data implementation and you have locked it into a legacy status.

- Standards based – In order for your service to work with other services, and work in other environments it needs to be based on accepted standards. If you build your service on some kind of custom build interface or home-grown web server, you won't be able to move it anywhere. If you aren't using standard HTTP/REST/WS-* semantics and deployments, it will be difficult to configure the web server to handle your services.

- Built with appropriate technology – I know this should go without saying, but empirical evidence suggests it needs to be said. You don't want to build a web service out of COM objects and Windows Services, and cron jobs. Stick with standard WS-*, Web API, SOAP style web services. If we are in a mobile app and services world, we need to communicate over the internet in most cases. At the very least we need to communicate over internet technologies even if they are VPN or direct connect. Use those technologies.

When we build out services, we need them to not only be portable, but modular. If you build massive monolithic services that encompass multiple domains, changing one part can mean redeploying the entire thing. That doesn't mean every service operation should be in its own service library. It means that you should group logically similar services and operations together.

Let's say you are building some mobile apps to handle all of the HR services for the BHD Company. Even though all of the services are related to the HR domain, there are logical subdivisions that make more natural modularity boundaries.

- Labor law compliance
- Recruitment
- Payroll and benefits
- Employee relations
- Employee performance

A large HR system could probably be deployed as five services, with the related operations for each discipline. They can certainly be named and deployed in a way where they seem to be part of the same service, for example:

- Labor law compliance – http://www.bhdcomp.com/hr/Compliance
- Recruitment – http://www.bhdcomp.com/hr/Recruitment
- Payroll and benefits – http://www.bhdcomp.com/hr/Compensation
- Employee relations – http://www.bhdcomp.com/hr/EmployeeRelations
- Employee performance – http://www.bhdcomp.com/hr/Performance

This deployment allows changes or replacement of the discipline-level components for the HR domain without affecting the entire HR service catalogue.

Another aspect that you can break services out into, is common application processes. This is similar to the ideas that Fowler [35] put forward for separating Domain Services and Application Logic Services. Consider the fact that you probably need services to do push notification processing or security processing. These will be common application processes across your entire mobile app space. Where you have these cross-cutting concerns, you can also logically group them into services. As with the domain-specific services above, you won't want to put all of these kinds of services into a single monolithic service. You'll want to break them up into natural modularity boundaries.

- Authentication
- Data validation
- Logging and auditing
- Monitoring
- Push notifications
- Message ingestion

These are only examples of the kinds of things you'll have in your application logic services, but it shows where the natural boundaries are. If you combine services like Authentication and Push Notifications, there is a discord that I imagine you can feel in your bones like fingernails on a chalkboard. (Does anyone still have chalkboards?)

While logically you may group Monitoring with Logging and Auditing, you will still want these to be separate service modules so they can be changed independently even though they very likely work in concert as in Figure 6.12 like a process improvement system. Monitoring will likely feed to Logging, and Auditing consumes the output of logging and suggests where more monitoring is needed. But the operations involved in each service are different and they could feasibly be used in isolation. You may need to monitor something for uptime and being

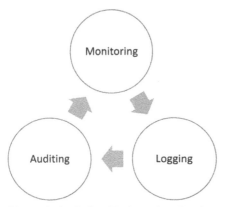

Figure 6.12   Related Independent Services

alive, but not log its activity. You may run audits on a system but check for current configuration and data usage without relying on the logs from the logging service.

The idea is that while the services work closely with each other, they are independent separate services that offer different functionality. They can each be used in isolation, or by other services without the other ones.

Making your services modular provides a lot of versatility and agility for your mobile app services. It lets you adapt quickly, and more importantly, swap components in and out when you need to. You may decide that your home grown logging system isn't all you need it to be and decide to adopt Enterprise Library or another third party logging service. If all of your services are built into one big service deployment, you can't do that without redeploying everything.

6.9.1 Designing Pluggable Services

One of the things that tends to get in the way of a good user experience is downtime. While downtime is a necessary evil in modern computing, it can be minimized if not seemingly eliminated (through the use of failover and redundant systems). Traditionally we've had to update systems to upgrade them, fix bugs through patches, or add new bits to the system. This usually required the admins to stay all night or spend the weekend updating and bouncing servers. In these cases, employees expected downtime, and were not usually surprised if Monday morning things were still a bit shaky. But that doesn't go over very well in a consumer-facing application, or a SaaS Service that people are paying for to be up 24/7.

To avoid this problem, the components of our system need to be hot swappable. That is, we need to be able to add, remove, or change them while the system is up and accepting traffic. In a traditional in-house data center, with everything running on the metal this is not easy to do. But in our new world of virtualized self-healing systems, it's quite possible.

The first component of this mix is the pluggable services. You need to be able to configure your services to be plugged in or removed while the system is operational. In most cases, this isn't just a feature of the service itself, but how it is deployed and most importantly how traffic to the service is being managed.

To be pluggable there are a few things that a service needs to be:

- **Stateless** – The service itself needs to either be stateless, or at the very least not keep state in the service itself. Now, keep in mind that the core advice of REST, and indeed HTTP is stateless servers. That is the spirit in which we should operate, but it's become quite common for systems to be developed that do indeed maintain state even with stateless technologies involved. Just like you do with a web farm, use an external session state server whenever your system requires user state to be maintained. Ideally, you'll be using REST and adhering to the HATEOAS principles which will eliminate this problem for you entirely.

- Messages to your service must be **Idempotent** – When you take a service out of the rotation, and plug a new version or patched version in, there is a good chance that at least one of the inbound messages will be processed twice. I talked about the idea of services being idempotent before. This is one of the reasons for that.

- **Configuration based** – The services that you plug in to your service layer need to pull their environmental information from configuration files. This not only allows them to read from a common configuration file for the deployment, but also allows them to be redeployed to a different environment and adapt at runtime. For example, you may deploy your service into a testing environment. In this environment the connection strings and authentication information are specific to the testing environment. Once the service passes all appropriate tests, you can then deploy it to the production environment where it will pull the production environment information.

- **Self-registering** – If you bring a new pluggable service online, it has to be able to register itself with the SIL. This can be done through Service Registries that are monitored by the SIL or the load balancer, or startup code that calls out to the SIL and identifies itself as available for message processing. In most cloud services, this is done either through scripting, or API calls. Keep in mind that this is not scaling out new instances of a service which most cloud providers orchestrate automatically. This is plugging in a new service that was not in the standard catalogue before. While this is called "self-registering" this can actually be done as part of the deployment of a new service through the deployment scripts if actual self-registration is not viable in our situation.

- **OS isolated** – The service also needs to isolate itself from the operating system. It's not that the services shouldn't talk to the OS, but it shouldn't perform any operations such as registry entries, or rely on OS environmental variables like changing the PATH statement, etc. that would cause the system to have to be rebooted. Or more importantly, that might not be there if the services are moved to a different environment that may not operate exactly the same. This is not an issue if the service is installed on its own VM, or into a cloud PaaS deployment because the service will be isolated from other services and the operating system anyway, but if you have to deploy into a shared VM, or physical web server you need to avoid anything that would require a system reboot.

These characteristics will provide you with the ability to drop a service into your service infrastructure and have it spin up and start taking calls without disrupting other systems, or requiring system reboots. This gives you hot swappable services that allow you to update your services without downtime.

The second component in the mix is the networking and traffic routing infrastructure. If adding a service requires you to reboot, recycle, or somehow disrupt the inbound traffic on the network it's not a pluggable service. The components that perform these operations might be the SIL components, or a traffic router, load balancer, or even a traffic manager service in your deployment.

Make sure that the service can register and unregister with the traffic management system without disrupting the traffic flow or popping messages off the queue and eating them without processing them. That is why the message timeout and retry are important.

6.9.2 Swapping Services

Now that you have modular pluggable services, how do you go about creating an environment that they can be hot swapped into? The process of swapping services is mostly how service traffic and monitoring are managed. There is also a difference in what exactly you are doing. You might be adding a new version of an old service, a new service, or changing one of your service for a third party service or vice versa or just removing a service.

Plugging in a new service is probably the easiest of the scenarios. You simply deploy the service, and then register that service with the appropriate networking and message passing infrastructure. (If you don't have self-registration in place) Once this is done it's ready to start accepting traffic and processing messages.

The other scenarios are dependent on the networking and message routing infrastructure in place. This is why we have an SIL. The SIL provides that layer of message routing. So the SIL needs to work in conjunction with your service swapping deployments. So the SIL is where most of the message routing components will reside.

In most large-scale internet-facing service deployments, you at least have multiple instances of a service running with a load balancer in front of them. This is the absolute minimum technology stack required to facilitate the removal of a service instance from a deployment. The issue though, is that it's not going to recognize new services without some kind of intervention.

Remember, that for all of their usefulness, load balancers are comparatively dumb components. They simply distribute incoming messages to each instance of the same service in the rotation of service instances. This isn't even among different services, just separate instances of the same service in a load balanced set. This tends to be the default for load balancers. They distribute the incoming requests to each receiver in the deployment in order. This is shown in Figure 6.13.

In some load balancers this can be changed to be more intelligent. These tend to go beyond typical load balancing and are often called message routers or traffic managers. They implement the Message Router pattern [34] which intelligently routes traffic based on CPU load, latency, or even server affinity rules. However, this does require a rule base to inform the Message Router how to decide which service or instance of a service to route the traffic to. This is a good thing though because it allows you to use a message router that can actually route messages to different services, even in different locations.

6.9.2.1 Service Removal Service removal, being the simplest process is just taking a service offline and decommissioning it. Message Routers tend to be required for gracefully swapping out services. This is usually done by telling the message router that an instance of a particular service, or the entire service is no longer accepting traffic. The message router will then take the service instance or service out of the destination list and stop sending traffic to it. Once the service instance or service completes the processing of its current message or job, it can then be shut down and removed from the deployment.

You don't want to just delete the service registration from the message router service registry. This tends to stop things pretty abruptly and exceptions start getting

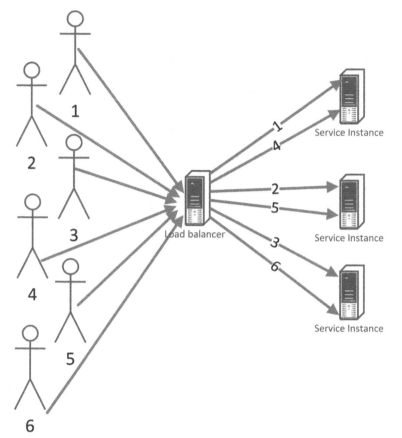

Figure 6.13   Round Robin Load Balancing

thrown. What you want to do is suspend the service from the rotation and let the message router return a more friendly response code than a 400 or 500 series error. Once the service is removed and the clients are updated so they know not to call that service anymore, you can remove the service registration entirely.

6.9.2.2 *Service Version Addition*   There are times when you will deploy a new service or a new version of an existing service. You might deploy the new version of an existing service alongside the old version for a while until you are certain that no one is using the old service, or you reach the end of the support lifecycle for the old service. At that point you can decommission the old version.

In either case, the deployment is the same. To deploy a new service/version of an existing service you simply deploy the service, and then register that service with the message router. In the case where you are deploying new version of an existing service, you will need a service versioning strategy which I will discuss in more detail in Chapter 9 Strategies for Ongoing Improvement below, but you can have multiple versions running side by side this way.

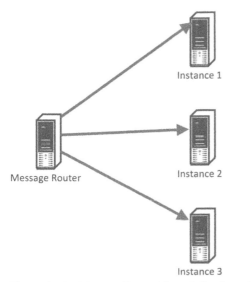

Figure 6.14   Message Router Normal Configuration

6.9.2.3 *Updating Existing Services*   Sometimes you'll have to patch or update your services. When you do that, you'll need to manage the traffic to the instances of the service while you do a rolling update. The idea is pretty easy to understand once you have a good message router in front of your service deployment.

You need to have multiple instances of the service running for failover and load balancing as in Figure 6.14.

You tell the message router to take Instance 1 out of the load balancing rotation. It will move the traffic over to Instances 2 and 3 as shown here in Figure 6.15. You then upgrade Instance 1.

Once Instance 1 is running and stable you add it back into the rotation as in Figure 6.16. Note that if your client apps are configured to ignore what they don't understand, they will be fine calling the updated service providing there were no breaking changes. If there are, you are in a multi-version scenario and not really just updating or fixing an existing major service version.

You repeat this process until all of the node instances have been updated and are back in the load balanced rotation shown in Figure 6.17. At which point the traffic will be evenly distributed between the instances.

Fortunately since this is a very repeatable process, most of this can be scripted or done through various management tools. I would suggest that if your environment can't be managed through scripting or some management tool suite, that you might have gone stray with your technology choices. Being able to manage this kind of thing through automation is critical to ensuring predictable results and eliminating human error.

6.9.2.4 *Changing from One Service to Another*   This scenario tends to happen during major technology shifts, or with the explosion of SaaS services that people

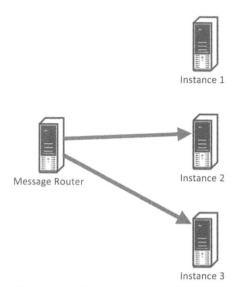

Figure 6.15   Instance One Out of Rotation and Updated

want to consume rather than building their own. In this scenario, you want to remove an existing service from the system, and either replace it with a different one either first or third party, or route the traffic to a different service.

This will require some interactions from the SIL as well for our message translation. Chances are the new service won't have exactly the same signature as the old service. So you will need to send your messages through the message translators, filters, enrichers, etc. Hopefully this will be pretty minimal if you have been following

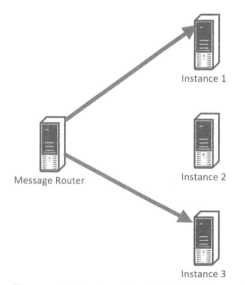

Figure 6.16   Instance Two Out of Rotation and Updated.

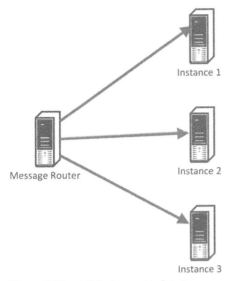

Figure 6.17   All Instances Updated

the Standards Based mantra I've mentioned throughout this book. Ideally you'll just have to reformat the message to get the right values into the right fields which is what we do in the translation components in the SIL.

The message routers though will need some consideration. When you were hosting the actual service in your deployment, your message router essentially handled the intelligent routing and in some cases load balancing for you. If all you required was a load balancer for your service and you are not using a message router, then the load balancer for this service can simply be decommissioned when the service is removed because that then becomes the responsibility of the new service provider.

What you do in this scenario is change your SIL components to route traffic through the translators to the new service, then decommission the old service when all traffic and processing has stopped on the nodes.

6.9.3 Deployment and Hosting Strategies

Service deployments are obviously critical to the success of your apps. Unless your app is a completely self-contained app on the device, in which case you wouldn't be reading this book, it will rely on your services being deployed in a high-availability manner. You will need at least some planning for High Availability/Disaster Recovery (HADR). You will also want to test your services in an environment that is as close to production as you can get to ensure accurate testing. To achieve these two goals you need to have in house IT resources, a hoster, or cloud service provider that can accommodate staging environments and automated deployment.

With on premises infrastructure, the cost of having true production mirrors is usually cost prohibitive for all but the largest organizations. Even with most hosting companies they will charge you for all of the infrastructure you use or have on stand-by for testing. Why not? After all if you have it on reserve they can't rent it to anyone else. Again, this is a costly scenario. Public cloud providers though tend to offer some level of this as part and parcel to using their services. You can even use trial accounts to experiment and do some initial feasibility testing. In either case, you want to be sure you have some infrastructure that closely matches your production deployment.

This is where those cloud providers tend to shine again. As I mentioned with a hoster, or on premises infrastructure, if the machines are there but not being used, you are still charged for them. With cloud providers, you only pay for what you are actually using but the capacity is there for you to use whenever you want it. When you are done with it, the CPU, RAM, storage all go back into the pool for someone else to use so you don't have to pay for it.

If you do use a hoster, ask them how fast they can provision and de-provision resources for you. If they only need a couple of hours, or if you can do it yourself through a control panel or scripting interface, you will be fine. If they need a couple of days to do this, you might want to consider a cloud provider.

Another aspect to consider that I mentioned earlier is the idea of PaaS versus IaaS. If you are a group of mostly developers, with little or no server administration experience, you will want to use as much PaaS as you can. PaaS eliminates most of the headaches associated with servers such as patching, image management, updating those images, etc. If you do have administration skills though, then it becomes a question of what are you more productive on, IaaS virtual machines, or PaaS services.

6.10 WRAPPING UP

This has been a marathon section. But it's a critical one in getting your services lined up so that they will carry your mobile apps into the future, whatever the industry may come up with. This means you have to be ready with pluggable service architectures deployed in highly scalable infrastructure. These high-volume services will be crucial to your success in the mobile market.

The system needs to be designed from the ground up with scalability in mind. Because scaling up is not a good practice with commodity infrastructure, we want to design systems that scale out. This means designing for server cluster environments. Plan for external state, avoid session affinity where possible, and build in good resilience so that we can scale out and scale back as required. Thinking in "nodes" will keep you on the right track.

For our services to be as efficient and resilient as possible, we looked at the CQRS pattern. This particular design pattern allows us to have dedicated read and write sides to our service. This allows us to optimize each side for its purpose so we can avoid locking data where possible and streamline authentication and message passing. When we combine this with Event Sourcing, we also get an instant repeatable audit train and recovery mechanism. CQRS also inherently allows us to be flexible in our data storage, message passing, and service interface technologies. It inherently allows us to swap, change, and route traffic using whatever technology is most appropriate for the situation now or in the future.

The services we design this way need to be as agile and pluggable as we can make them. With mobile apps and services, there is this unwritten expectation that they will be available 24/7. So downtime, even minimal as it may be, is a big deal. With pluggable service architectures, we can avoid downtime during updates and the addition of new services. Designing these kinds of pluggable services takes some planning up front to ensure we chose technologies that can be automated as much as possible, and that our potential hosting platform can support that automation.

Creating a robust service layer is where the bulk of the work is. It becomes the power behind your apps, and it is where your real value add is. This is core of your system. Spend the time to lay it out properly, and you'll be well set for the future.

CHAPTER 7

THE DATA ABSTRACTION LAYER

7.1 OBJECTS TO DATA

We write services that operate on objects. They work with defined contextual constructs in memory that represent our real world entities. These are often not in the format that we need them to be in for permanent storage. We need to be able to translate them into something that can be written to disk in many different forms such as Relational Databases, NoSQL Databases, and Flat Files. That is what the Data Abstraction Layer does.

In other circles this is often referred to as a Data Access Layer. While this term suits the purpose, it is not encompassing enough to capture what has to happen in this layer. There are additional functions this layer performs that are not just about accessing the data. This layer also translates, transforms, and does some level of communication routing. It provides an abstraction between the physical data storage, structure and location, and the services layer. Because of this, I prefer the term Data Abstraction Layer, as it more accurately describes what is happening here. Of course, if you see the acronym DAL, and in your head you hear Data Access Layer, it's not like the world will end.

The data abstraction layer is basically a layer that isolates the services layer from the underlying physical data storage mechanisms. This prevents tight coupling, as well as a tendency to put too much service logic in the database. This is often built using a Data Mapper Pattern [35]. Although modern software has outgrown the original Data Mapper pattern as described by Fowler [35], it is still quite applicable and provides a good way to separate domain from physical data store. Sometimes these layers are Object Relational Mapping (ORM) systems such as Hibernate ORM, Entity Framework, or some other object to data storage mapping system which implement this pattern.

ORMs and similar technologies combine many patterns to facilitate separating domain objects from database level storage. They also use a Repository Pattern [35] to hold the object collections in memory after retrieving from the database. In modern mobile app scenarios, we would also include various message translation and routing patterns as well. We are well beyond the time when a system was only connecting to a single data source, and NoSQL and Services are as much a part of system data as any relational database.

Designing Platform Independent Mobile Apps and Services, First Edition. Rocky Heckman.
© 2016 the IEEE Computer Society, Inc. Published 2016 by John Wiley & Sons, Inc.

The important part about this layer is that it is responsible for the translation and routing of data from the Service Layer to the Data Layer. The translation is from in-memory contextualized object representations, or other code-level constructs to some kind of serialization or calls to a database, or potentially another service run by your organization or a third party.

This layer can be as complex or as simple as required. For example, if you use an ORM it has the ability to do this conversion for you and may do everything you need. You just configure the ORM for the data store you are using and call its methods from your code. Different ORMs have different capabilities, some only work with relational databases, while others work with just about any physical data storage mechanism you could use. For example, there is also Hibernate Object Grid Mapper (OGM) which gives you Java Persistence (JPA) support for saving your Java objects in NoSQL data stores.

ORMs can greatly simplify keeping the database data structures and the code in sync. Most of them offer some form of data schema update function when the apps and services are deployed. In Entity Framework (EF), for example, this is called Code First Development. There are couple of aspects to this that are quite interesting. During new development, you can have EF create the new database and build out the schema for you. This is called Code First for a New Database. You can also have it insert the new data model for your classes into an existing database. This approach actually reverse engineers the database and create skeleton code from the schema found there. This is predictably called Code First from Database. This can be very useful if you already have data in place that you need to plug in to.

In both cases, when you make updates to your class structure, attributes, properties, etc. the Code First Migrations can be used to keep the database schema in sync with the data model from your codebase. All of this is well documented on the Entity Framework website: https://msdn.microsoft.com/en-us/data/ee712907#codefirst

Now, having sung the praises of ORMs, they are not without their dangers. They do introduce several orders of complexity to the system. You will also find that once you start using an ORM, you are quite tightly locked-in to that ORM. With this complexity comes performance issues. I've often told ISVs and customers I work with that there is this tradeoff between development effort and end-user simplicity.

> *Rule of Thumb: The more effort the developers put in, the easier things are for the end users.*

The same holds true for using things like ORMs. Some clever middleware developers have created these things to simplify life for developers that are doing the same thing over and over. That is persist code-level objects in permanent storage. This code chunk is repeated in every major application. Why not have a way to plug in some library that can not only handle that for you, but also keep your objects and storage formats in sync? So enter the ORMs. We need to stop and think about what they are doing though.

These ORMs are doing their level best to create a physical data model that can be rehydrated into the logical data model in the code and vice versa. Well, in order for them to be as data source agnostic as possible, they could not assume certain vendor level mechanisms would be there so they effectively hand-craft a lot of the SQL in a database or dynamic SQL in the code, for example. This takes a considerable amount of code to get right. There are no less than 46 namespaces [56] across 12 components that make up the 8–9 layers of ADO.NET Entity Framework [57]. Messages going through all of that stuff get slowed down.

Part of this performance problem is because they must be designed for the lowest common denominator across all databases and storage mechanisms. EF, for example, generates "appallingly bad SQL" [58] that does not perform and isn't scalable enough for most internet-scale implementations.

ORMs also lag behind, sometimes considerably, in data type support in underlying data stores. An example of this is, as long as we are on the EF bandwagon, in SQL Server 2008 we had Spatial data types. This wasn't introduced in EF until version 5 which was introduced in August of 2012 [59].

So you have to tread carefully around ORMs. They offer great simplicity of use, and keep the object model and the data model, such as it is, in sync. You will have to deal with performance and scalability trade-offs though. Sure they are great for avoiding vendor DB lock-in, but as I mentioned in the beginning, at some point, to get the most out of a back-end system, you need to take advantage of the vendor-specific features of the platform you've chosen.

This doesn't mean you have to tightly couple the Service Layer to the Data Layer. But you can write your own DAL that performs just the operations of a typical ORM or OGM that you actually need. Plus, you can fine tune those operations for your domain model. This gives you a lot of flexibility and control over the DAL operations. This also gives you the ability to quickly adapt to persisting data with services, rather than local data stores.

With this context in mind, it is important to remember that existing ORMs such as Hibernate and EF are not mutually exclusive with direct or facilitated (JDBC, ADO.NET, etc.) calls to the database. You can use an ORM as your primary method for creating and manipulating individual documents or records. Even sets of them, but when it comes to updating many records at once, such as setting all of the "expiry dates" or archived flags, you can certainly call a stored procedure to update any or all of the records. You don't want to do things like that through an ORM because it means you have to fetch, load, change, save, and persist every single document or record. That is wildly inefficient. It's easier to call an "UpdateExpiryDate" stored procedure and have it touch all the affected records. So using a combination of these two technologies is the most versatile and efficient approach.

Both of these approaches should be implemented in your DAL. Mass data operations should be represented as a method on the factory or data collection that calls through to the stored procedure. While small group or individual record manipulation can be done through the standard ORM/OGM mechanisms.

To get an optimized DAL you will need to line these two things up together. This means that you will need to implement some level of the patterns mentioned above for your data access. You might even decide to go so far as to create a customize ORM that is optimized for your organization. This has certainly worked very well for Facebook and Twitter. You can also consider factors such as with your domain objects and choice of data store, if you are using the CQRS pattern, which is more optimized for your read and write sides?

7.2 USING THE DAL WITH EXTERNAL SERVICES

Typically ORMs and OGMs do not translate and route persistence request to other internal or third party services. You will have to implement translation and routing components for these kinds of services much like the ones listed in Section 5.5. In fact, the physical implementations of these translation and routing components may be deployed in the same location as the SIL components. Remember that the five layer architecture is a logical architecture, and physical deployment can vary.

The logical flow of communications from the UI Layer down to the Data Layer goes through the SIL, to the SL, and the DAL to the DL, respectively. This is depicted in Figure 7.1.

This may not be the actual deployment. In the physical deployment, the flow to any third party persistence mechanisms from the DAL will probably need to be translated and sent to a service endpoint that is outside your physical deployment. When this happens, it is the same kind of communications that you have for services

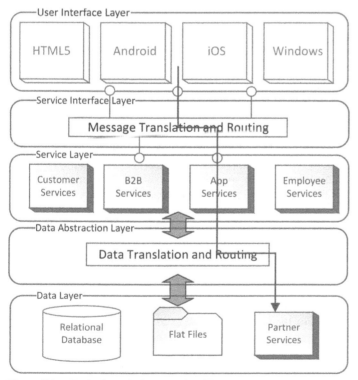

Figure 7.1   Typical Logical Interaction Flow

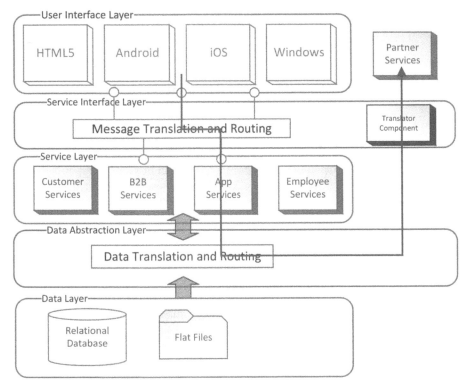

Figure 7.2   Physical Interaction Flow

that interact with yours through the SIL. A visualization of this would be more like Figure 7.2.

Since both the DAL and SIL perform routing and translation it may at first be confusing to see them both doing the translation of the message. Remember the DAL is there for your services to write their higher level objects to, then the DAL translates them for physical storage. If you are sending information to an internal first party, or external third party, there are different levels of translation you may have to use. Remember the diagram in Figure 5.9, it showed that there are different levels of translation.

The DAL may need to convert the code-level object to say JSON for storage, or into SQL commands for storage, but the DAL by itself is not where you do the data format translation for service calls. The DAL may perform the data structure and even data type translations, but the SIL is where all of your Data Format translations will occur except in special circumstances.

The DAL is used to translate from code-level data representations to storage level representations. Typically the DAL won't need to do Data Type and Data Format translations because it's storing information in physical mechanisms that are designed for your use anyway. So the data types and data format will already be known and quite likely already in the proper formats. Because the DAL doesn't deal in WS-*, SOAP, REST, or HTTP, that portion of the translation is reserved for the SIL.

Following our flexibility mantra, should you decide later to stop using an externally hosted third party service and deploy that service directly into your infrastructure, or implement a different persistence mechanism all together, you just change the endpoint that the DAL is sending the persistence commands to. This may require that the translation component be swapped out, changed, or potentially removed entirely depending on your choice of new Data Layer storage mechanism. But you will essentially stop using the SIL unless you are persisting the data in an internal service.

7.3 COMPONENTS OF A DAL

So our DAL is produced with three main components:

- An ORM/OGM or your custom code to data structure mapper
- The module that helps with bulk updating of records and data, where using the ORM would be too inefficient
- The router that sends calls to third party services through the SIL

The ORM/OGM components are pretty self-contained and you will likely use an existing one such as Hibernate or EF rather than re-inventing an entire ORM. That is, unless you want to create a highly optimized one. So this is a pretty easy component to figure out. A message router that routes the persistence message through the SIL to an external service isn't too hard either. The custom component for bulk updates and optimized ORM (OORM) style persistence is a bit more challenging though, so I want to go over this in a bit more detail.

When we think about this component, it can be one of, or the only component in the DAL. In some cases it may just be passing on calls to a bulk update stored procedure, in others it may be performing customized serialization, aggregation, and distribution of data elements. In either case though, it is one of the pieces of your system that has to understand the object model, and at least some of the data storage model.

What you end up doing is implementing several architectural patterns in a very lean, specialized ORM. You will have at a minimum a data mapper and a query mapper. Ideally you will also implement some form of object caching probably in the form of a repository. This repository will also be responsible for tracking changes to the single instance of the object and persisting changes back to the database.

Your bulk CRUD operations will be performed through the query mapper. This mapper may simply translate object field names to database column names while performing the query, or may be more complex to the point of performing many joins, name translations, and reference data access. The complexity is dictated by how well your domain objects align with your database tables. Keep in mind that all of these mechanisms can also be built to work with NoSQL data stores. So a query mapper may actually be replaced with simple Document Database style storage calls. In these cases, the custom ORM will take the business object, serialize it as a JSON document, and then pass it through the query manager to be stored in the Document Database.

Let's briefly look at some of the patterns involved in creating an OORM.

7.3.1 Data Mapper

At its heart the DAL is a data mapper. Its primary job is to translate from domain objects to data storage calls. It is there to provide a layer of separation between the Service Layer and the Data Layer to prevent the SL from being too tightly coupled to the DL. The DAL itself also performs other operations, but this is where data mappers sit.

The idea behind it is that when the SL requires an object to be instantiated from the data store, it calls the DAL and connects with a data mapper. It then asks the data mapper to find the object it is looking for. For example, if we want to get an instance of a Blue Hair Dye company customer, we would ask the data mapper for a Customer with a particular ID. The call may look like GetCustomer (12);

At first glance it appears that this would simply map to a database call to a sp_GetCustomer stored procedure. But there is more to a data mapper. What should actually happen is that the data mapper checks an entity map to see if this Customer had already been retrieved and is already being used in the system. It should return that mapped instance and monitor who has an instance and track updates. This last bit is tricky. You have to track the last update and then know if a subsequent update is working on a stale version of the object. So you will need to implement a data versioning strategy in the entity map.

You will also have to decide what level you are going to map your entities on. At the lower end of the spectrum you can have an entity map for each table or record store in the data store. You could move this up a level to store them at the logical entity level. For example, if you have an Order object, you can store it complete with its line items instead of an entity map for the order, and a separate entity map for the line items. Even higher up the scale is say a Customer, complete with their orders, and associated line items.

You need to be careful how high you go though. Data Mapping at a high level, say AllCustomers means a lot of nested tracking and huge data loading operations. You have to consider where your "Get" methods will be aimed. If you more frequently get orders, than whole customers, and hardly ever get line items by themselves, then perhaps having a Data Mapper at the order level, and a separate one for customers is the best approach. Your Data Mapper can put the two together when required. For example, if you had two separate UI operations, one that worked with just customer data, and one that worked with the customer's orders, the first would primarily call the GetCustomer operation which would cause the Data Mapper to use the Customer Entity Map. The second one would usually call the GetOrdersForCustomer operation which would typically cause the Data Mapper to use the Orders Entity Map. Knowing these typical use cases will help you optimize your entity maps to make the Data Mapper more efficient. Your Data Mapper will contain the appropriate Entity Maps for the level it is designed for. An Orders Data Map, for example, will likely have Entity Maps for Orders and Line Items in it.

The above cases were examples of an Explicit Entity Map. Each operation was specific to a particular entity, and hence an entity map. This is an efficient and verifiable way to build out a set of Entity Maps. However, if you want to be more flexible, and you know you'll be adding to, or changing, the Entity Maps over time, you may want to implement a more generic "Get" method over your Entity Maps.

This would look more like Get (Customer, 12). In this case, the type of entity that is being requested is passed into a top level generic method. That method then looks up the correct map and pulls the correct entity. This is easier to add, remove, and change things with. There is a drawback though, you lose some of the strong typing checks you can do at compile time that you have with calling and returning a known entity type from something like GetCustomer(12). You may have to implement Generics to accommodate this.

Another aspect you want to consider to make your Data Mapper as efficient as possible is to separate the immutable entities from the editable ones. If you have loaded entities that you know you will never have to change such as reference data, keys, etc. then you don't have to bother tracking changes or doing data versioning. The more code you can eliminate, the more performance you are likely to see. So consider breaking up your Entity Maps into immutable and non-immutable categories.

Immutable Entity Maps can be used from anywhere in the system and need to have that visibility level. Any of your SL calls should be able to quickly see any of the read-only reference data from a single location. For example, you may decide to physically implement your immutable Entity Maps in something like a Redis Cache accessible by all the instances of your system.

On the flip side, non-immutable instances of entities are normally associated with a particular session context. A sales person will likely need their particular set of customers loaded into their session without wanting to wait for all the non-related customers to load. Additionally, you want to keep track of different copies of the data as they relate to the end user. For example, your list of products can be loaded into an immutable Entity Map, but the users shopping cart and list of items in it, with their customizations such as size and color, need to be loaded into the non-immutable session contextual Entity Maps.

A word about instantiating objects in the Data Mapper. Since the Data Mapper is largely responsible for hydrating the objects from the underlying data, you need to consider how you build out your objects and how to deal with internal only fields, etc. Normally you'll want to use explicit constructors and populate them with the data that may be read only when the object is used. You can create empty objects and then set all the fields, but that forces you to have setters for everything you would need to set when the object is hydrated.

7.3.2 Query Mapper

Query Mappers (QM) are essentially a way to convert object syntax into physical data storage syntax. In some cases, this will be translating it to SQL, in others it maybe to a JSON Document DB, flat file, or external service call. QMs are used in place of an ORM sometimes to represent complex object hierarchies or nested objects. Sometimes if you are retrieving data from separate data sources on the back-end, or you need to create some conditional nested constructs, and ORM may not be able to

Figure 7.3   Customer Class

deal with this very well. In these cases you may need to hand-craft a QM to deal with complex data assembly.

In any case, QM will have to keep track of incoming objects and field names, and outgoing targets and data names. For example, if you have a Customer object, and you want all Customers with a last name of Vader and our structure looks like Figure 7.3.

You could do this a couple of ways. You can create a Get method that is something like GetCustomerByLastname("Vader") or pass in a Customer Object and have the QM pull as much information as possible from it to find the match. You might use reflection to do this, or some form of a dependency injection.

What you end up with if you use reflection for example is something like the following:

```
private static Customer GetCustomer(Customer cust)
{
    StringBuilder where = new StringBuilder();
    SqlConnection conn = new SqlConnection
(YourConnectionString);
    SqlCommand cmd = new SqlCommand("GetCustomer", conn);
    cmd.CommandType = CommandType.StoredProcedure;

    foreach (PropertyInfo propertyInfo in cust.GetType().
GetProperties())
```

```
    {
        if (propertyInfo.CanRead && property-
Info.GetValue(cust, null) != null)
        {
            // Map your Property Name and Value to your Data
Store Name here.
            cmd.Parameters.Add("@" CustomerMapper.
GetDataNameFromObjectPropertyName(propertyInfo.Name),
propertyInfo.GetValue(cust, null).ToString());
        }
    }
    SqlDataReader rdr = cmd.ExecuteReader();
    rdr.Read();
    // get the results of each column
    string firstName =
(string)rdr[CustomerMapper.GetDataNameFromObjectPropertyName
("FirstName")];
    string lastname =
(string)rdr[CustomerMapper.GetDataNameFromObjectPropertyName
("LastName")];
    DateTime birthDate =
(DateTime)rdr[CustomerMapper.GetDataNameFromObjectProperty
Name("BirthDate")];
    int id =
(int)rdr[CustomerMapper.GetDataNameFromObjectPropertyName
("ID")];
    rdr.Close();

    return new Customer(firstName, lastname, birthDate, id);
}
```

In this example, we use reflection to identify the properties we can access from the object. We start off here with an advantage because we know we're dealing with a Customer object. This is perfectly fine and you may have a Query Mapper for each object type in your system. Because of this, we know which properties we have to work with, and what we'll have to pass to the constructor at the end to return a valid Customer object.

In this example, we use a Metadata Mapper to help our Query Mapper create a set of database query parameters. We do this by iterating over the properties of the object, and for each one that was presented to us that is not null, we add a parameter to the list of Query parameters with a corresponding stored procedure parameter name. We do this instead of ordinal values because if the object, or the database changes, ordinal values will break.

TABLE 7.1   Simple Metadata Mapping

Object Property	Data Column
FirstName	fname
LastName	lname
BirthDate	bdate
ID	Id

As you can see the Metadata Mapper also helps us avoid having to know anything about the names of the parameters in the database, and the database doesn't have to know anything about the names of the object properties.

If we wanted to do a bit more work, we could make this more generic. We could accept an initial parameter of type Object. Then get its type and its associated properties. This would mean fewer Query Mappers, but it would mean a much more complex single query mapper, and a more complex metadata mapper. We could create a metadata mapper for each object type, and map the property names and data column names. Alternatively we can create a single large metadata mapper, and have a way to define which object and property we are trying to match a data column to.

A simple metadata mapper may map like the one shown here in Table 7.1.

If we were to have a more generic query mapper, we would need to be able to identify each object and its associated properties through metadata mapping. We can either create a metadata mapper for each object, or we can create one metadata mapper and design the name storage to identify the object.property names and the table.column names. We can store the map using an Object.Property to Table.Column pairing similar to Table 7.2.

This second method can be used to map all of our object.properties to table.columns or whatever our data store is using. This is very similar to the service parameter mapping that is done in the service registry for the SIL. They operate in a very similar manner. The Message Translators and Message Routers use the Service Registry in the same way that the Query Mapper uses the Metadata Mappers.

Query mappers can be as simple or as complex as required. This is one of those components that the more development effort you put into it, the easier it is to develop against it later. The DAL is one of the critical layers in future proofing your services from changes in the data layer. Inevitably things will change in the data layer. More and more people are moving away from traditional relational databases and this means changes to the data layer and the way you access the data stored there.

TABLE 7.2   Type and Property Metadata Mapping

Object Property	Data Column
Customer.FirstName	customer.fname
Customer.LastName	customer.lname
Customer.BirthDate	customer.bdate
Customer.ID	customer.Id

The query mapper insulates the SL from changes in the DL. Especially for the components that are making fairly direct data layer calls. Remember that these are things that are not going through an ORM or external service. The QM will work with bulk operations or loads and saves that the ORM may be less efficient at. You will probably want to set up some bulk operation methods in the QM that allow for mass updates in the data layer such as zeroing out stock levels for inventory, or touching active records. In these cases your QM will pretty much just pass through the call and format it for the target data store.

While a QM is certainly logically associated with a relational database, you can use it in conjunction with a Repository, as a way to assemble data from many different sources some of which will likely not be relational databases as I mentioned earlier. For example, you may have an Orders table in a relational database that uses line items from a NoSQL table store like MongoDb or Azure Table Storage, with product information from a NoSQL Document Database like Azure's Document DB or Amazon's Dynamo DB. In either case, standard query pass through to SQL won't get the job done. Your QM will need to pull information from those two or three different sources and aggregate it into an order object as shown in Figure 7.4.

While a QM from its nomenclature would not normally be considered this way, in today's world of ever changing data stores, we need to adapt. If you can't use an ORM for performance reasons or bulk update reasons, then your QM will have to be able to handle these kinds of operations on whatever data store you have in your DL.

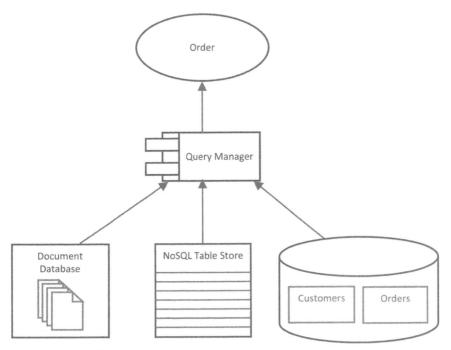

Figure 7.4   QM Data Aggregation

But try to keep it limited to the bulk operations and doing things a standard ORM can't handle. It's easy to add too many things to the QM and pretty much duplicate an ORM.

If your QM starts to look more like an ORM, you might want to consider if you should just be using an existing ORM for that portion. Keep the QM lean and make sure it's providing efficiency, not just duplicating all the features of an ORM. You may just end up using a QM for forming appropriate queries and passing it to the DL and use an ORM for the more robust objects. An important aspect of a QM is to start small, and add as you need to.

7.3.3 Repository

Repositories can be thought of as an extension of Query Mappers. They handle things slightly differently. Instead of supplying methods like GetCustomer(int id), a repository allows calls to specify a set of criteria to match, and the repository produces one or more objects that meet those criteria. In essence, a Repository is a Query Mapper combined with a Metadata Mapper, and access to a caching mechanism.

They have fairly generic methods which return one or more objects based on the supplied criteria. There's a chance that you've been thinking of the non-ORM portions of the DAL as a Repository to this point. You're probably right. If we combine the patterns we've talked about so far with a Cache, we have a Repository.

You can have methods such as GetSingle(TypeOf(Customer), CriteriaCollection) which identify what they are looking for based on the first parameter, and apply the filters based on the criteria such as FirstName = "Darth" or LastName = "Vader." In some cases, the object type is combined into the criteria itself, for example

```
Repository.GetSingle(new Criteria(typeof(Customer),
"FirstName", "Darth", CriteriaType.Equals));
```

The method above will return an object which should be of type Customer. This very generic method allows us to call it with just the information we want, and reuse it for just about any object in our system. Obviously the internals of this method will need to be able to accommodate pulling the information from whatever data sources, and our cache, that it needs in order to find and return the object we are looking for.

Similar methods need to be created for GetMultiple, GetAll, etc. Be cautious with GetAll methods though. You may end up pulling thousands or even millions of records from the DL, and that will definitely be expensive. You need to have proper paging in place if you plan to do that.

As you can see, the Repository is very similar to a Query Mapper combined with Metadata Mapping, Caching, and an index of already pulled objects. You'll want to be sure and not hit the DL unless you really need to, while balancing this against the possibility of stale data and data changes either in the cache of the Repository itself, or in DL data store.

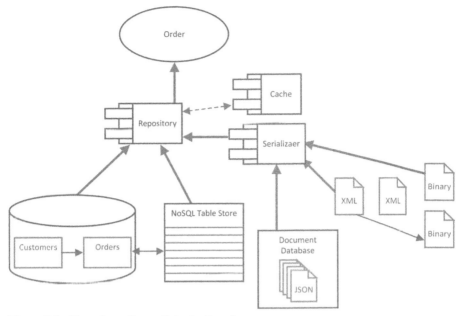

Figure 7.5   Repository Over a Polyglot Data Layer

Keep in mind that with all of these things form QMs to Repositories that they may have to pull data from multiple different data sources including external or third party data sources. In fact, your custom QM/Repositories will more likely have that job than would a typical ORM.

When you have multiple back-end data stores of different types, the Repository Pattern is probably the best way to go when it comes to accessing all of the different data stores. It allows you to present a uniform interface to the SL while abstracting away the individual nuances of accessing data in the DL from a myriad of different data stores. In our orders example, you can see by the diagram Figure 7.5 that the repository provides a nice way to collect the data for presentation to the SL.

In this situation, we have customers and orders stored in a relational database, line items stored in a NoSQL key/value store, JSON documents representing the product catalog information in a Document Database, and serialized XML and binary files used for productivity documents like spreadsheets and text documents as well as embedded images, or custom binary files we've stored. While this may not be typical, it illustrates the point. The Repository will be able to collect all of this information and cache it where possible. This way we can present something sensible to the SL, and not have to deal with tightly coupling the SL with a myriad of data storage technologies.

If we ever add, remove, move, or re-arrange this data layer deployment, we can make those adjustments in the DAL/Repository and the SL keeps on ticking without knowing or caring what happened on the back-end.

Some people have asked me, "so why don't we just do all of that in the service layer? It seems like just extra work". This may be true at first glance, but remember how services are accessed and who is accessing them. We need to maintain that lose coupling. Remember, lose coupling is the key to future proofing our apps and services. If we had to make changes to the service layer to accommodate any moves in the data layer, it means we need to redeploy the service each time we make a data layer change. That means potential downtime or errors bubbling up to the client.

By making these kinds of changes in the DAL, we can bring up new configurations, or adapt to changes in the data layer but keep the services running calling the same endpoints and using the cache while the change propagates. Any potential errors we can deal with internally in the DAL, and the ripples in consistency, or error propagation won't reach the clients.

7.3.4 Serializers

One of the things that the DAL has to do is prepare data for storage. Aside from breaking things up into SQL query parameters, storing things a bit more naturally is also part of its job. To this end we have to serialize objects into one form or another for storage or transmission to external services. There are three commonly used types of serialization, Binary, JSON, and XML. We talked a bit about serialization in Section 2.2.4, as it relates to preparing data for transmission. We'll cover it here in a bit more detail.

A note here, we aren't going to talk about all of the possible serialization formats. We are only going to focus on the ones that are applicable to modern mobile apps and services. So we aren't going to cover things like YAML (YAML Ain't Markup Language, previously Yet Another Markup Language), which while it is a superset of JSON hasn't really worked out for broad adoption. We also won't really talk about the Property List format as it pretty much used within the Apple OSX eco system and not very useful for cross-platform, platform-independent apps and services.

Serialization is the act of converting an object or structure in a computer program and flattening it into a reversible string of bits so that it can be stored or transmitted much like the Transporter in Star Trek [60]. Data can then be easily handled for storage or transmission in a manner that is decipherable by the reader of the serialized data. The reversible part is important. If it can't be reversed you've effectively lost your data. More importantly in our context is that it must be reversible by other systems that we may not have written or have control over. This is why JSON and XML serialization have been the defacto web standards to this point.

JSON and XML are text-based serialization formats. They are popular due to their universally readable format. Binary serialization, while typically more efficient, is normally not readable without very specific knowledge of how the object was serialized and the objects structure prior to it being serialized. So you only really use binary serialization when you are in control of both the serializer, and the deserializer on both ends of the process.

For all of your transmissions internal to your system, or even between programs on the same system, using Binary serialization is a good choice. It is very easy to transfer through memory-based channels, or directly to other services running on the same machine. It can even be sent over the wire usually as hexadecimal encoded strings of characters. Since you know exactly what it is you are going to be reading, you can use binary internally across the board if you like. But when you need to send serialized data to some app or system outside of your complete control, you'll need to use JSON or XML serialization.

With the rise of Document Databases, serializing object to JSON format has gone from common to default. If you are using any kind of document database, you will inevitably serialize your objects as JSON. JSON is also one of the most common formats for transmitting data over the wire to web services and mobile devices. It has pre-built libraries that support over 60 different programming languages [61]. JSON is an ECMA standard (ECMA-404) [23] and has very wide adoption. This is largely due to its simplicity and ease of readability. A testament to its simplicity is the ECMA standard itself, the entire thing is only 4 pages long. (and it's mostly pictures.)

The JSON format is also the most common format for RESTful web services. RESTful web services also use ATOM formats, but they aren't as popular as JSON formats. In fact most commonly used RESTful web services on the Internet don't even bother to supply an ATOM format. This is entirely up to you of course.

JSON serialization tends to be as efficient as it gets as far as text formats go. It eliminates all of the extra information and types, or custom classes that formats like Binary and even XML offer. It is very simple in that there are essentially five things you can find in a JSON serialized document; object, array, value, string, and number.

XML Serialization on the other hand offers the ability to do custom types and create very complex serialization through various schema definitions. It's this ability to define custom schemas that allows you ultimate flexibility in the serialization of your objects if you use XML. These schema definitions can also be a huge source of pain if you reference the wrong one, or forget to reference it altogether. Even with these issues, XML is the most popular text format in use today. Many traditional web services have been using XML for many years. The default transmission format for WS* SOAP-based web services is XML. We have even seen the SGML-based XML format become pervasive in file serialization.

Most modern document and office productivity suites such as Microsoft Office, Open Office, and Apple iWork products save documents in an XML format. In the case of Microsoft Office, the Open XML format became an ISO standard (ISO/IEC 29500) after it had been standardized by the ECMA as ECMA-376. So when we think of our DAL serializing objects and documents for storage, we need to consider our target audience.

When we talked about serialization in Section 2.2.4, we were thinking primarily of conversion and transmission to and from web services, mostly in our message translators and web services. But our DAL is concerned with persisting data into the DL for storage either locally or remotely or even in an external system. So we need to consider how we address serialization for storage in our DAL.

7.3.5 Storage Consideration

When we think of our DAL and its job, we need to consider what our target storage mechanism is, and who is likely to need to read data from that storage mechanism other than us. Many times over the years people have developed systems thinking that their code was the only code that would ever read it. Only later to discover that all kinds of other programmers wanted to access it, and third party systems, and eventually someone slapped a web service interface from some company on there and the world wanted to read the data. Back then, this was a travesty of translation.

Most of the time, the data were stored in very proprietary formats. Even to the point of storage on mainframes, VAXes, and other super computers. You were only getting to that data if you were a COBOL expert with a ponytail and a huge coffee mug. The best part is, that was the way the world wanted it. Now, it's all backwards. We need to design data to be accessible first, and proprietary hardly ever.

Let's look at what we are trying to do first and our target system. Chances are if we are at the point of considering a serialization technique, we aren't talking about relational databases. We are in the realm of external services, flat files, and NoSQL stores. While you can certainly serialize an object and put it into a database column, if you are doing that you are thinking of a NoSQL Documents Database, not a relational database.

Our format will in part be dictated by our target storage mechanism. As I just mentioned, if we are working with a NoSQL Document Database, then we are going to be serializing objects or documents into JSON and storing in the DB. This is the most common method in the current NoSQL Document Databases like MongoDB, Amazon's Dynamo DB, and Microsoft Azure's DocumentDB.

However, if we are working with or storing productivity documents such as spreadsheets, presentations, and well, documents, then we should probably consider XML-based formats such as OpenXML which every productivity suite in common use today can read and write to.

If we are writing out information to a file, and our system is the only one that will read it, then binary serialization is the way to go. While this persistence format will be very efficient, it also means that our system is likely the only one that will be able to use it, but in this context, that's fine because we are after all a DAL.

Because the DALs job is to translate from objects to data storage and back, as long as all access to the data layer goes through the DAL, we can serialize information in whatever format we want. If, however, we may be shipping the files, documents, or other serialized data to other systems, we'll have to be considerate of that and stick with standard, likely text-based, storage formats.

There is a bit of a guide on the ISOCPP.org site on serialization (https://isocpp.org/wiki/faq/serialization) they had a pretty good breakdown of how you choose between text-based and binary serialization. Here is a quote from their site:

How do you choose between binary and text serialization.

- **Text format** is easier to "desk check." That means you won't have to write extra tools to debug the input and output; you can open the serialized output with a text editor to see if it looks right.
- **Binary format** typically uses fewer CPU cycles. However, that is relevant only if your application is CPU bound and you intend to do serialization and/or unserialization on an inner loop/bottleneck. Remember: 90% of the CPU time is spent in 10% of the code, which means there won't be any practical performance benefit unless your "CPU meter" is pegged at 100%, and your

serialization and/or unserialization code is consuming a healthy portion of that 100%.

- **Text format** lets you ignore programming issues like sizeof and little-endian versus big-endian.

- **Binary format** lets you ignore separations between adjacent values, since many values have fixed lengths.

- **Text format** can produce smaller results when most numbers are small and when you need to textually encode binary results, for example, uuencode or Base64.

- **Binary format** can produce smaller results when most numbers are large or when you don't need to textually encode binary results [60].

It is quite a good starting point for deciding Binary or Text.

As with all things in our realm of mobile app services, it's not an all or nothing gambit. You can certainly binary serialize files that are of a proprietary nature, or that are standard binary formats such as images, sound, and video files. The documents that reference those files, or that may have had those elements embed in them can be serialized into JSON and stored in a Documents DB with pointers to the binary files. Even information stored in a relational database can have pointers to binary files or references to JSON documents stored in a NoSQL Document DB.

Let's take our example from earlier where our Query Manager was assembling information from several sources to return an object to the SL. We'll use the diagram below as our talking point.

In Figure 7.6, we can see that there is a complex relationship amongst our serialized data. Customers will be associated with one or more orders, but the order can be made up of several different pieces of data that have been serialized and stored into their most appropriate and efficient formats. While some may say you can put the whole Order object into a Document Database, that doesn't always work so well with very large items or items where pieces such as line items and products need to be shared across other documents. But I digress.

An order is made up of customer information, and several line items and embedded documents such as customer instructions or artwork. Each line item is made up of order-specific information like quantity, and other information such as product description, pictures, and even instructions. The catalog item information is stored in a Document Database, and has links to binary files such as product images and instructional videos for the product. We store this here because our Catalog Generation system also uses this data on demand to generate custom catalogs based on a user's location and available products in their region.

As you can see, we end up with a polyglot data storage system. Multiple serialization types for multiple purposes in the most appropriate data storage mechanisms. While this seems complex, it offers us that flexibility and future proofing we need for rapidly changing mobile apps and services.

When we have to retrieve a complex item such as an order, our Query Manager or Repository will have to work with the Serializer to deserialize the various documents and objects where required. Other components in the DAL may need to

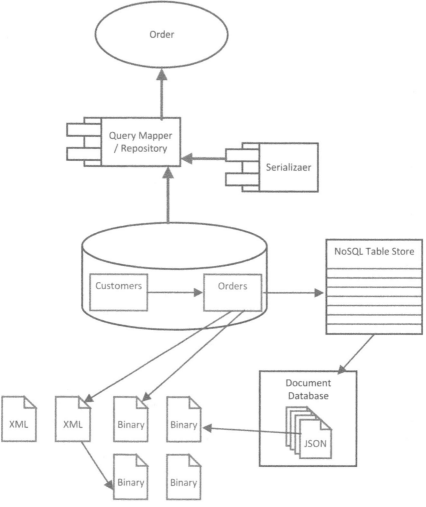

Figure 7.6   Polyglot Serialized Data

serialize objects for storage as well. So creating a central serialization component will be worth the time due to the broad reuse it will get.

7.3.6 Cache

Caching data is critical to performance. You will want to have some level of caching in the DAL. Unlike the UI layer cache, the DAL cache will allow you to cache method calls and returns. This is similar to how a relational database stores the execution plans of recent queries and their result sets. Your DAL just does it at a higher level and in a more contextual manner.

The Data Mapper's Entity Maps act like a cache of sorts. Indeed they are a cache for hydrated objects, but those are not designed for streamline bulk operations. In fact your Data Mapper will likely use the DAL cache as a way of making its database calls more efficient where possible.

There is a balancing act, here you will have to observe between DB caching and DAL caching. In most cases where your physical data layer is a relational database, you will rely on the DBs internal caching mechanisms to handle query result caching. But in cases where you are using a NoSQL style of data store, or flat files, you'll need to rely on the DAL cache.

Normal caching rules apply here. Don't cache data that has security restrictions on it. If the data updates frequently, caching the data may be a waste of time. Cache data that do not change frequently such as reference data, or static BLOB data.

Something else that I've seen a lot of people do is try to use caching to overcome poor data access and database query design. I was working for one organization that offered teleconferencing services. One of the database queries we had to deal with was literally 42 pages of 12-pt font long. It took several minutes to run over moderate data. It basically built a customer's bill for our services. This is not something that can be overcome with caching.

There are several places that developers use caching in an application or back-end system. Most people think of caching in front-end services and web pages such as web server caches for static pages, or caching the results of a jsp or aspx page to prevent it from being rebuilt. At this layer we are more concerned with data caching.

There are many guides and best practices around using application data caching. Most of them use some form of framework cache class, or some *ad hoc* key/value structure like a dictionary object. These do work, and can help, but they are not designed to do caching of data for presentation to multiple calling systems.

The DAL needs a caching mechanism that can service the needs of the SL without resorting to mechanisms that tie it into one memory space, or application context. So for caching in the DAL, we need to consider external cache mechanisms or writing your own cache component that wraps these other mechanisms.

As it happens, NoSQL data stores of the key/value pair variety have become quite popular as external caching mechanisms due to their speed and extreme scalability. There are many open source and commercial varieties available. The more notable open source ones are: Redis Cache, MongoDB, couchBase, and memcached. Some of the commercial ones include XAP, BigMemory, and Pivotal GemFire. You can also use cloud-based technologies such as Microsoft Azure Tale Storage and Amazon's SimpleDB as a cache if you chose. For a DAL implementation, things like Redis tend to work very well. While other open source NoSQL document databases will work fine, Redis has been tuned to act more like a cache, than some of the others.

The critical thing is that it needs to scale and perform very well. There's no point in having a cache that takes longer to return data than calling the original source. This tends to be assisted through the ability to partition your cache and keep the lookups very fast. That being said, Couchbase is introducing multi-dimensional scaling which will increase their already leading performance marks. Couchbase may be overkill if we are just looking for a caching solution though.

Your cache must be able to store serialized data, as well as objects and binary files. Consider this and what you use most when choosing your caching technology.

You may cache more serialized documents and fewer query returns or vice versa. You'll need to evaluate your specific scenario to find the best solution for you.

You also want to consider if you need a security layer on your cache or not. If you can guarantee that you won't be exposing your cache to anyone or anything outside of your control then you might be able to use generic security credentials for the cache or disable authorization checking altogether. However, if your situation is such that you may have other developers, systems, or partners accessing your DAL for various reasons, you'll want to tighten down the security aspects of the cache.

Some people say that this will hurt performance. No, it really won't. Or to be more specific, authorization checks are less than a rounding error when it comes to performance in an overall data retrieval process. If you are having performance problems, and you are at the point of considering turning off authorization checks to get 0.00001 seconds of response time, then you need to back up and re-think your data access strategy. It's more likely that you'll need to consolidate some of your data storage mechanisms, or perhaps partition your data better.

The cache will be the first port of call for your Query Mappers and Repositories. Most ORMs come with their own built-in caching. So you'll want it to be organized and you'll want to make sure you keep it cleaned up. I have seen people build systems that never cleared anything from the cache, and they tried to cache everything. So what happens is that the FIFO rules apply.

Things that were cached a while ago have to get pushed out to make room for new cached items. A good cache will apply some intelligence here and remove the items with the oldest touched time. Some don't. Cleaning up your cache makes the cache more responsive because the cache has to do it at the time it goes to insert the next item that slows down the insert. If you keep it tidy, and your cache never has to free up space during an insert, it will respond better. So cache judiciously, and tidy up regularly.

7.4 WRAPPING UP

When we get down to it, the DAL needs to be able to handle a polyglot data storage space. This means that it has to be able to facilitate data storage and retrieval to a very mixed bag of data sources, some examples of which are shown in Figure 7.7. Relational databases, NoSQL key/values, document databases, graph databases, and flat file storage. It has the additional task of having to route communications to other services external to your system. This is a complex thing to accomplish. But it's critical to ensure your service back-end is able to move with the times, and always take advantage of the best technology for your growing needs.

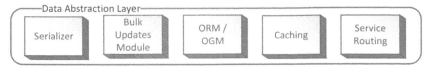

Figure 7.7   DAL Components

The DAL is one of the components that has to understand your logical models and your physical models. It provides that abstraction layer between the SL and the DL while facilitating smooth and efficient access to objects for your SL. You will spend a considerable amount of effort here. You might decide to implement just an ORM, and stick everything into a relational DB. That is certainly a time-tested and well-known approach, but as internet scale systems such as Facebook, Twitter, and Instagram have shown, the NoSQL approach provides excellent performance and scale benefits.

CHAPTER 8

THE DATA LAYER

DATA **IS KING**! The data you collect, store, manipulate, and provide are the core of what you are doing. An app that doesn't deal with data at some level probably isn't doing too much. Even an app that just says 'Hi' when started up then closes has to hold the greeting somewhere at least. Any apps that you want to remember anything about the user need to store it somewhere. So you may store a little data, or you may store so much data that you need a full data warehouse implementation.

Data serve a purpose beyond simply recording what happened and playing it back as well. Data can be mined, analyzed, and charted to give you impressive insights into how and where your app is used, as well as insight into the people using it. Some companies such as Google and Facebook have staked their entire business model on collecting, analyzing, and selling data about their users to advertisers and other third parties.

To get this level of value from your data, you need to store it logically, and in a manner that allows you to efficiently collect, process, retrieve, and analyze it. This is why data storage mechanisms and data science are a staple of computing and have been since the beginning.

The point of a data layer is to have a permanent storage mechanism to store data that we will work with later. The data layer is perhaps the most well-established layer and technology we use. This is the durable storage mechanisms you use. Essentially, it is what you do now with data. There are three basic ways we store data, relational databases, NoSQL databases, and flat files as shown in Figure 8.1. You need to pick the type or types that are right for you. Keep in mind that it's usually a combination of these, rather than all the eggs in one proverbial basket.

You need to focus on how the data will be used from its stored form. Are you just going to pick up an image and send it to a web page? Do you need to find relationships between various pieces of data? Perhaps you just want a fast way to store loosely related items and retrieve them very quickly. We'll discuss these factors and what you need to think about in this chapter. Specifically we'll look at the three main areas of relational databases, NoSQL databases and file-based storage. Then we'll consider how these all fit together in a polyglot data storage system.

Designing Platform Independent Mobile Apps and Services, First Edition. Rocky Heckman.
© 2016 the IEEE Computer Society, Inc. Published 2016 by John Wiley & Sons, Inc.

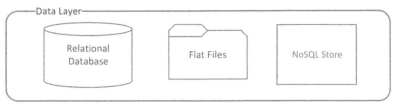

Figure 8.1   Basic Data Storage Mechanisms

8.1 OVERVIEW

Data layers are becoming much more NoSQL centric. The simplicity of storing and retrieving documents is what is attractive about this if you can just set up a document store and tell the system to stash documents in there, then you can search. And you can retrieve those documents at any time based on any information in that document. That makes things a lot simpler if your system is all about information storage and retrieval. However, this technology doesn't lend itself well to analytics, business intelligence, or big data scenarios from and information extrapolation. That doesn't preclude you from being able to do this level of data mining though. Things like Map reduce, jobs and Hadoop allow you to scan and gather information about these kinds of unstructured data sources. Everything from twitter feeds/sentiment to full information extrapolation and prediction are possible.

So why bother with relational databases at all? Primarily because you can work with normalized data. For example, in a NoSQL store you can have several documents from the same person all with the same name and address on them, but if that person changed their address, you don't have an easy way to update that address on all of those documents. You have to either leave them as they are and map that change to all historical transactions, or you need to cycle through all of the affected documents and update them. Neither of these approaches is very attractive. When it comes to keeping live information on customers, systems, etc. relational databases are going to offer you more flexibility and ease of use. But if you are just storing documents in a static fashion, and you need good search and retrieval of them, then a NoSQL store is going to suit the situation a bit better. If you don't need to update the information in the store, then NoSQL is a great option, but if you are regularly doing updates to the information in the store relational databases are a better option.

The data layer in the architecture is designed so that you can use whatever data store suits the type of data you are working with, without having to worry about changing the layers above it if you change data stores. A lot of systems now, especially line of business systems, are built on top of relational databases. Some of them are even exposing web services directly from the relational database itself. While this is a really easy way to surface access to the data through web services, it is a serious technology lock-in and tightly couples your services layer to your data layer. This is counter to what we need to accomplish with future proof mobile app services.

8.2 BUSINESS RULES IN THE DATA LAYER

Many line of business systems have implemented a lot of their business rules in the database itself. Usually these are implemented through complex stored procedures or customer functions. While this is a good idea from a performance perspective, and it guarantees that access even at the data layer will enforce these rules, the problem is that this tends to tie that database to a particular system. There will quite likely be times when you need to expose that data to a different system or to other external partner systems. If you have enforced these rules in the data layer, it restricts the value other systems can get from the data layer. Don't get me wrong, those rules are very valuable to your systems, and the performance improvements are nice, but that isn't where the only value is.

Business logic, rules, decision, all of it should be done in the service layer. The only kind of rule enforcement that should be done in the data layer itself is security and data integrity checking. These will apply across any system that interacts with yours, even ones that you haven't thought of yet. You'll need that kind of flexibility to ensure that you don't have to recode the internals of the database each time a new service comes along, or you integrate with a new partner. This is especially true if you decide to change physical data stores. If you move from say a relational database to NoSQL store, or to a third party service you'll have to re-write your entire business rules layer into the service layer anyway. This kind of rewrite cannot only blow out the cost of the project, but it can also scrap it entirely if it's too much trouble to do.

8.3 RELATIONAL DATABASES

Relational Databases are perhaps the most common enterprise data store in use today. Everything goes into a database so it can be analyzed, prodded, poked, and massaged to provide some new insight into business and our ever accelerating industry. That's fine though, because relational databases are really good at that sort of thing. They can even be designed to optimized data storage and retrieval, as well as enforcing data validation, or even some business rules if you have no other way to do it.

One of the biggest advantages to relational databases is of course the relationships that can be built for the data. For example, a user has a relationship with one or more addresses. It's these relationships that allow you to avoid data duplication. If you stored the user information complete with address information in one row in a table as illustrated in Table 8.1, if they had three addresses, home, work, and billing, then you would duplicate the user information in each of those rows.

TABLE 8.1   Denormalized Address Table

ID	Alias	First	Last	Street	City	State	Country	Zip
007	RockyH	Rocky	Heckman	123 Main Street	Springfield	NT	USA	80085
007	RockyH	Rocky	Heckman	124 Main Street	Springfield	NT	USA	80085
007	RockyH	Rocky	Heckman	122 Main Street	Springfield	NT	USA	80085

TABLE 8.2   Denormalized Phone Number Table

ID	Alias	First	Last	AreaCode	Prefix	Line
007	RockyH	Rocky	Heckman	636	555	5566
007	RockyH	Rocky	Heckman	636	555	6655
007	RockyH	Rocky	Heckman	939	555	8008

TABLE 8.3   Normalized Address Table

Id	UserId	Street	City	State	Country	Zip
110	007	123 Main Street	Springfield	NT	USA	80085
111	007	124 Main Street	Springfield	NT	USA	80085
112	007	122 Main Street	Springfield	NT	USA	80085

TABLE 8.4   Normalized Phone Number Table

Id	UserId	AreaCode	Prefix	Line
21	007	636	555	5566
22	007	636	555	6655
23	007	939	555	8008

Then what if we had a phone number table done the same way as in Table 8.2.

So what happens if they change their Alias? You have to change it in three rows in two separate tables: Table 8.3 and Table 8.4. But with the magic of relationships and a process called normalization you extract the duplicated data into its own table, and put a foreign key constraint on the other tables that link that information to the single record of truth in Table 8.5.

You can even take this one step further. Many people may have the same work address. So you might only enter each address once in a table, then put a table between the UserInfo table and the Address table that links the two together by containing two foreign keys, one to the UserInfo table keyed on UserId, and one to the Address table keyed to the AddressId.

This also allows for very useful things like cascading update and cascading deletes. For example, if a user needs to be removed from your system, you simply delete the user record, and the database cascades that delete to all of the records in all of the other tables that contain a foreign key to that user record. This is a very handy way to keep the database clean, consistent, and error free. I recommend using foreign key constraints in your relational databases to prevent things from getting really scrambled if something goes wrong if you try to do it from code. The database

TABLE 8.5   Normalized User Info Table

Id	Alias	First	Last
007	RockyH	Rocky	Heckman

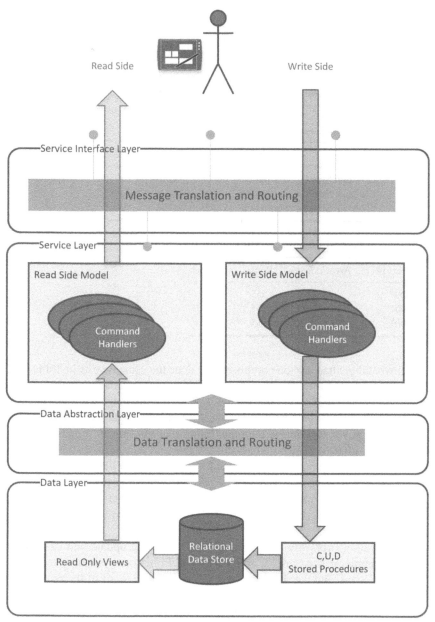

Figure 8.2   CQRS/ES with a Single Relational Database

will always get it right if you set up the foreign key constraints. A code level bug can scramble some data or leave orphaned data in tables. Worst case scenario—this could introduce a data exposure security bug.

In the CQRS/ES pattern, a single relational database can serve as both write side and read side data stores as shown in Figure 8.2. The write side can use standard insert operations to insert records into the database. The read side can use read-only Views or Stored Procedures to read the data. While this is still an eventual consistency model, it prevents you from having to move data from the Event Source to the read side source. All you do is change the access model, and enforce rules that only allow the insert, upsert, or delete through a specific set of stored procedures only callable by write side services, and that all read operations go through a similar set of optimized services authorized for the read views and stored procedures.

8.4 NOSQL DATABASES

NoSQL Databases have become quite popular recently. In fact large organizations like Facebook used them as we mentioned earlier. Don't' let the moniker get in the way though, some of them do actually use a Structured Query Language (SQL) syntax. What people mean by NoSQL would be more accurately called Non-Relational databases. They are databases though in the sense that they are an organized collection of data. [62] They were original devised because relational databases didn't have the large-scale performance, and were often too structured for the simple tasks that people wanted to perform such as long-term storage of key-value pairs, or storing serialized documents.

There are basically four types of NoSQL databases, key-value stores, document databases, column family databases (CFDBs), and graphs. You may find one or more of these quite useful in high-performance mobile app services. This is especially attractive, given the simple interactions of these kinds of data structures.

There are even SaaS services now for the various NoSQL database types such as storing and searching JSON data directly such as Amazon DynamoDB and Microsoft Azure DocumentDB. With systems like this, the JSON documents are stored unaltered and their fields can be searched through various SQL, Language-Integrated Query (LINQ), and object methods. This represents a very high-level storage tier. You don't have to call a stored procedure to store the data, and you don't have to know SQL to get the information back out and re-create the objects.

However, updating data in these types of systems means that the objects that the documents represent have to be rehydrated in code, their values changed, and then saved back to the NoSQL store. When compared to a relational database where changes can be made at the row level, this is wildly inefficient. So as tight as your code may be and as "internet scale" as the NoSQL store might be, it's still not going to perform at the level you need if you are storing dynamic data and doing regular updates to it.

Another thing to consider is that traditionally, these types of data stores have not been real good at transactional operations. Additionally many of them suffer from reliability problems. NoSQL is used widely in large social network systems. Anyone that has used those systems has at some point gotten an error saying "something went wrong try again later". If you are ok with losing data once in a while, and in some cases it's perfectly acceptable, then this may be fine.

Where NoSQL excels is in rapid document storage and retrieval. For example, if you are doing CQRS/ES write-side operations for accepting PDFs or JSON style documents, or even file uploads, this would be a great way to store the documents or links to the file storage locations. You can even combine this with the write side to read side push and persist data into a relational store if you need to do updates on the metadata, or say user, name, address type fields and use the NoSQL store as the static store for the document itself.

NoSQL stores also tend to be highly scalable due to the limited amount of massaging of the data they do. They are great at accepting massive amounts of inserts, and very good as high-scale reads. They are designed to store and retrieve lots of data as quickly and simply as possible. They do it very well.

One of the things you'll have to consider regardless of which NoSQL stores you use, is how to key the data. Rapid retrieval is based on good partitioning with unique keys. This depends on if you are using one or two part keys. Some of the key-value storage ones use two part keys, a partition key and a row key. The data are grouped by primary key and then the row key is used to identify the data row. This allows the data to be spread out into shards and makes it much faster to process because the table doesn't get very deep.

If you are using one part keys, then each one has to be unique and it's difficult to shard the database, unless the database generates a simple hash of the key and groups by that hash. Note that I am not talking about genuine MD5 or SHA 256 hashes here, but a simple algorithm that produces values from a small and finite list. There are many ways to figure this out, you can either key the row on a value unique to the data in the row, such as e-mail address for a customer, or you can generate the key.

There are obviously several ways to generate the key. You can use a concatenation method, where you simply do Firstname + Lastname + DateTime + something else. Alternatively you can use GUIDs. GUIDs are always unique and should be collision free. But again, they are static and need to be remembered somewhere for lookups. The data value or concatenation method works pretty well. Ideally your data will be designed in such a way that you can use logical identifiers as the partition key such as Region, State, Company, etc. and then a unique value for each row. Sometimes, this isn't so easy. You might not be able to easily identify a partition and row key. You may only have one unique value in your data, or you may require three, or more, pieces of information to uniquely identify a row. In the first case you might need to use a hashing algorithm. The information can also be used to produce a partition key for partitioning.

For example, let's say we use a hashing algorithm that takes our row key, which is the e-mail address of the user, then takes its length and does a mod 5 operation on it. That will produce the numbers 0–4. That becomes our partition key in which case we'll have five partitions and the rows within the partition will be keyed on the e-mail address. Table 8.6 shows how this would work.

In this case, we end up with five partitions with each containing a section of the list. While this distribution seems wildly uneven, it is a small sample and the laws of averages would work in our favor the more records we had. Obviously if you only had ten entries, this would be serious overkill.

Alternatively if you have three identifiers, you might need to use the highest level as a table identifier, the second grouping as a partition key and the last one as the unique value. If you start going beyond three pieces of information to uniquely

TABLE 8.6   Hashed Partition Key

Row Key	Partition Key
Joe.Smith@bhd.com	2
John.Franklin@bhd.com	1
Sue.Blue@bhd.com	1
Alice.Springs@bhd.com	1
Bob.Baker@bhd.com	2
Charleston.Clarion@bhd.com	1
Dan.De@bhd.com	4
Ed.E@bhd.com	2
Frank.Frezin@bhd.com	0
Geoff.Geoff@bhd.com	4

identify a row, you need to re-think your data strategy and if NoSQL is appropriate. You can always fall back on multi-part keys, and then parse them. But if you end up with something too convoluted, you might be going after the data from the wrong side.

Let's say you run a huge machine parts store, everything from motors to nuts and bolts, from jet engine blades to cotter pins. You want to organize your product shelving. So you come up with this schema for your keys. Building + Floor + Section + Aisle + Row + Shelf can be used as your location information. You could concatenate this together as:

Table: Building
Partition: Floor
Row: Section:Aisle:Row:Shelf

You'll have to parse the row to get more details. But why would you do this? Chances are you are going to look up the location of the parts given a part number, rather than have a location to figure out what's in that spot. So really, you would want a categorization structure. Part numbers might not be unique, but chances are they would be unique within the context of a given manufacturer. So you should key your data on Manufacturer: Part Number and store the location in the data portion of the row.

This will be the same for pretty much all NoSQL data stores. You need to consider a keying mechanism that provides logical grouping for partitioning and uniqueness for individual data items. Choose your keys on whatever makes sense for the domain context.

8.4.1 Key Value Database

Of all of the NoSQL DB types, key value databases (KVDB) tend to be the best choice if the data update frequently. While you still can't do mass operations on data like you can with a relational database, the speed and single row retrieval in properly partitioned key-value store tends to be acceptable for looping if you don't have to do mass updates very often.

KVDB are the most basic way you can store data for later retrieval. You have a key that you will need to remember (keep in mind the key discussion we had above). You can normally store different types of data in the column sections of the database and it can be different from row to row. The data stored in the value portion of a KVDB is opaque to the KVDB itself. So you can't search on data values, or index it by data values. This is why the partitioning and key hashing algorithms of a KVDB are so important to its performance.

KVDB tend to be very fast if you partition them correctly. Most KVDB enforce a keying structure that allows or creates a partitioning scheme that is used to maintain that speed and efficiency. In many cases they either use a two or three part key, or use some kind of hashing function to generate a partitioning key and use a supplied row key. The efficiency of this hashing function can be a factor in the speed of the database. If you are tempted to build your own from various dictionary objects or something, make sure that you are very efficient with your partitioning logic, and how you create your hashes to determine your partition and data location.

For speed most KVDBs will use the hash as indicator of the partition, and a location within the partition to store the data. The hash values are stored in order to make look ups and retrieval faster. But you will run into a problem if your keys start hashing to common value and you get hash collisions. This means that two keys will hash to the same value, so the KVDB will try to insert the data in a location that already has data in it.

When this happens, it has to decide how to find a different spot to put the data. In some cases, it just slides down the rows until it finds an unoccupied one and puts the data there. In others, it uses a second hashing function to create a new hash based on the first one and then puts the data in the newly identified location.

Neither of these situations is ideal. This means that when you go to retrieve the data, the KVDB has to make sure that the data it found at the location the hash indicated match the supplied key. If it doesn't, the system has to presume a collision occurred on insert, and then follow whatever procedure it used to find a new location for the data on insert, and go find it there on retrieval.

This results in a point at which your hash functions will start to cascade. What happens if you have three key values that have hash collisions? If you are using the "find next available row" pattern, then your third data will be inserted two rows down from where it should have been. Table 8.7 illustrates this.

TABLE 8.7   KVDB Collisions

Key/Value Store		
10	AABBO	Nothing here yet
11	AABBD	First Insert
12	CCDF1	Collision 1 data
13	DFGAA	Collision 2 data
14	DFGAC	Other data
15		Nothing here yet
16	FFAC3	More data

Let's say we go to insert data and our hashing algorithm produces the hash that targets row 11. So we insert our First Insert data there. Then later we try to insert "Collision 1 data" and the key also hashes to row 11. Well we can't put the data there so we have to find the next open slot which happened to be row 12. Then we try to insert "Collision 2 data" and the results of hashing its key also targets row 11. Well we have to go to the next row, but it's full, so we end up finding space in row 13 and put the data there.

When the system tries to retrieve "Collision 2 data" the hashing algorithm tells it to check row 11. So it does, but we can't just return the data, we have to check to see if the key at row 11 is the key we are actually looking for. Since it's not, we now have to go down every row checking each key until we find the one we were sent to get. This turns into a linear search.

What happens when we have new data that hashes to row 12? Well since we took that slot with data destined for row 11, we have to slide down the rows until we find an empty spot to put it in, which in this case would be slot 15. This quickly becomes inefficient.

In essence, by having a very small hash result space, we've made the act of finding a location by the hash, somewhat useless. Other than determining a starting location to search for the data, it still turns into a linear search which is what we were trying to avoid. So you need to ensure that your keys are chosen well, and that the hashing algorithm your chosen KVDB uses has enough hashing key space to avoid collisions for the data set numbers you are going to store.

If you only plan to have 150 records in the KVDB at any one time, then you only need a hash that can produce more than 150 unique values. On the other hand, if you are going to store millions of records in your KVDB, you need a partitioning and hashing scheme that can deal with that and avoid collisions as much as possible. Some KVDBs let you plug in your own hashing algorithm. You might want to check on that if it's applicable to your situation.

These various collision strategies also create problems for data deletion. You can't simply delete the data in a row and make the row available. If you do, then the next retrieval attempt for data that previously had a collision will fail. In Table 8.7, if we deleted the "First Insert" data, then row 11 would be empty and available for use. But if we tried to retrieve either "Collision 1 data" or "Collision 2 data" the hashing algorithm would first direct us to row 11 to find our data. The system would then not find anything so it wouldn't know to follow the collision protocol to look for data in subsequent rows (or other locations if you used a second hashing strategy instead of next available location strategy). This results in all of the data that previously collided being inaccessible.

What KVDBs tend to do to avoid this is mark data as deleted, while leaving it in place. That way, if the system tried to retrieve "First Insert" data, it would report that it is not available because it was deleted. When the system tried to retrieve "Collision 2 data" it would find row 11 occupied, check the key, find it's not the one it's looking for, and then go through the collision protocol to find the data we're looking for.

This does produce a problem though. With lots of deleted data, you don't gain any efficiencies. Even if you deleted all but one row, you'd still have a lot of data taking up space, and the hashing algorithm would still produce a lot of collisions.

Periodically the database has to be compressed and cleaned up. This usually results in deleted rows being cleared, and all of the hashes recomputed to move the data back around to as close to its original hash location as possible. It's a pretty expensive operation, but doable.

As long as you've chosen a good KVDB, and there is plenty of key space in the hashing algorithm, KVDBs are fast efficient storage mechanisms that are great for basic storage and retrieval operations. They don't do ATOMIC updates so they aren't appropriate where you need that field level update capability. That's fine though, they weren't designed for that. They were designed to be very fast storage and retrieval of whole chunks of data. KVDBs are highly scalable and super-fast. They get their speed by using an eventually consistent approach and sacrificing some ATOMIC operation capability. Some of them such as Microsoft Azure Table Storage do offer a highly consistent model by getting write success acknowledgments from the participating servers. You'll need to check your KVDB to see what features it offers and what it doesn't.

8.4.2 Document Database

Document databases are designed to store serialized data in whole parts as a collection of name:value pairs. The value stored in the database is the serialized document itself. This makes them very suitable for things like whole pages of data such as an items catalog page, or a person's resume, or serialized objects. In fact with most of them the data being stored can be anything that can be represented as a collection of name:value pairs. A typical JSON file is a perfect example and many Document DBs are tuned to store JSON documents.

One of the reasons it is this way is because unlike a KVDB, the contents of the value being stored in a Document DB are visible to the database. Not only can a user query on the values of the fields in the document, but they can also create indexes on them for fast search and retrieval based on the indexed fields. Understandably, this means that most Document DBs required the document to be stored as something like JSON, BSON, or XML, something it can read the name:value pairs out of.

Documents are stored and retrieved based on a key for the document. This is very much like the KVDB in that the key is handle to the document. You will need a good key strategy as we've discussed before. Collisions in a Document DB are similar to a KVDB but because indexes can be created on the values inside a document, it's not as big a deal as it is with a KVDB.

Document DBs do use a partitioning approach to data, and in most cases you can define the partitioning approach. They also shard the database to provide optimal efficiency. Each Document collection or table has its own sharding and partitioning in place. Each shard typically contains one document table or type. Within that shard the partitions are defined to help make each shard as efficient as possible.

How you shard and partition your document DB should be dictated by how the data will be accessed and how the documents can be keyed and indexed. You want to ensure that the most common queries across the documents are served best. Your sharding and partitioning strategy should reflect that. You may want to see if you can customize the sharding process to keep related documents together. For example, if you have a Document DB holding Order documents, you may want to ensure that the sharding process shards based on Customer ID to keep all of a customer's order documents on the same shard.

Your shard keys should have a lot of available values, be highly random and most importantly reflect the most common use of queries. This will ensure that the shards are evenly accessed and that you avoid data hotspots while keeping similar data together for efficiency.

When you use a Document DB, one of the primary factors in deciding what and how to store the documents is in how they will be retrieved. Instead of storing everything as a document, and every document, consider the types of documents and their content for each Document DB. You need to keep like documents with like documents otherwise the indexes you create will only apply to a subset of the documents being stored. You will lose the efficiencies of the system by lumping everything together in one DB.

Keep in mind that a Document DB does not deal with relationships between entitles it is storing. So if you want to store Customer documents and Order documents in the same Document DB, you aren't going to be able to perform a JOIN like in a relational DB and pull all of the documents for a given customer by retrieving the customer document. You can certainly create an index on the CustomerID field in the order document, and pull all documents that have the same CustomerID field, but you can't do that by pulling the Customer document.

This also means that there is little in the way of understanding foreign key relationships. While you can do it by referencing other documents, there is no concept of referential integrity. You can't do things like cascading updates or deletes. Don't count on being able to delete a Customer and having all of their orders removed as well.

Think of it this way, if this was in a relational database, there would be only one table full of denormalized data. Everything you want to retrieve can be pulled from the table by one simplistic query with no inner or outer joins, or multi-table select statements. In a document DB there is one table, and it's full of documents. The more similar those documents are, the more efficient the retrievals are. So structure your documents, and indexes, based on whatever the service layer will be asking for the most. You should only need one trip to the Document DB to get what you want and the query is just a set of filters.

The other thing to avoid is the temptation to just make really big documents that contain every bit of information about everything. Most Document DBs have size restrictions on the documents. Aside from this, you aren't looking at the right tool for that job. If you are even able to put all the information about your domain entities into a single denormalized structure, maybe you should be considering a CFDB instead.

Since you store and retrieve whole documents in a Document DB, this makes them unsuitable for highly transactional workloads. If you want to update a field in a document you have to pull the entire document, change the information, and then save it back to the DB. If you are doing this for a mobile app, over the users' mobile data plan, you are not going to be on their holiday card list.

Document DBs are best suited to documents that do not change much, if at all. A good use of a Document DB is collections of historical or record keeping type documents. Think of things like candidate resumes, filled orders, catalog pages, transcripts, signed legal documents, etc. Remember they are not meant for live transactional data.

You also want to avoid nesting unless there's a good reason for it. For example, you don't want to nest orders within a Customer document. For starters, this will quickly make the document too large to be stored in the Document DB. Also, you shouldn't be forced to retrieve a customer, with all of their orders, just to check what was on order number 3321. The flip side of this is that you don't want to embed customer information in an order document either.

Some people have told me that they need the customer's current address on the orders when they retrieve them. Really? An order that has been filled was sent to a particular customer at their particular address at the time the order was filled. If you change the address when the customer moves on all of their old orders, you have broken your historical records and probably a few laws in the process. What would happen if you get audited and they look at Imperial Tie Academy's order from 5 years ago? Maybe 3 years ago Imperial Tie Academy moved from Nantucket to Sacramento. So at the time of the order you calculated shipping from your Los Angles hub. The auditor in the black cape and shiny helmet is going to ask why the shipping was so overpriced to Sacramento. This also prevents you from having to go and update 500 documents if someone changes their phone number.

Documents in a Document DB need to be considered as they were when they were stored. I would consider a Document DB to be more like a searchable filing cabinet (organized by document type) rather than a live view of the world as it exists right now.

If you are working with data that update or change regularly you need to consider your strategy carefully. Let's say you are working with stock data. You may have a document for each listed company. But you don't want to put their stock prices and their market changes into the same document as their company information. You could implement this by putting the per-minute stock prices and market value into a separate Document DB or Table with those documents.

This would allow you to get a stock history by retrieving all of the stock and market value documents from the stock price table based on the ticker symbol rather than retrieving the whole company document and trying to pull the stock prices out of it. If those prices were embedded in the document, eventually it would be too large for the Document DB, and each update, which includes a full document retrieval, update and full document save would get slower and slower.

In reality, you would probably consider a KVDB to store something like stock prices, or anything that is being tallied on regular intervals. But we'll discuss blended strategies later in this chapter.

The other area where a document DB has initial challenges is with relationship modeling. Normally if you want to focus on the relationships, you're going to use a Graph DB. But there are occasions when you do need some level of relationship tracking in a Document DB. The most common way to do this is to make one of the fields in the document link to the one or many other documents that are part of the relationship as shown in Table 8.8.

TABLE 8.8   Single Document Links

Key (UserId)	Document
10	"firstName": "Alice", "lastName": "Apple", "mother": 3, "father": 4
11	"firstName": "Bob", "lastName": "Apple", "mother": 7, "father": 8
12	"firstName": "Charlie", "lastName": "Apple", "mother": 10, "father": 11
13	"firstName": "Debbie", "lastName": "Apple", "mother": 10, "father": 11

There are a couple forms of hierarchies or relationships you can model in a field. There's a typical foreign key type relationship where you can name things like parents or managers.

This is the simplest link to another record, a single value to another document key. If you want to show a hierarchy, you can use multiple keys in a field as seen in the value in Table 8.9.

In this case, Debbie reports to Charlie, who reports to Bob, who reports to Alice, who reports to whomever is saved in record 1. Ideally, this is a fairly stable company where people don't change jobs and managers very often. Given the nature of Document DBs, you don't want to be updating this on a daily basis. If you are doing a lot of relationship shuffling, consider the Graph DB.

Features of Document DBs vary quite a bit from product to product. Some of them allow Upsert operations (that is on document update, if it already exists it is UPdated, if it didn't it is inSERTed), some allow document locking, some enable projections and views. Some allow custom indexes while others allow you to tweak their indexing scheme. You really have to check out the options available to you and see what will fit your needs the best.

8.4.3 Column Family Databases

If we want to optimize a bit further, and deal with data in logical chunks rather than a series of key/value pairs, we can look at Column Family (sometimes called Column Oriented) databases. As we have gone from simple KVDBs, where we have a key and a chunk or chunks of data, to Document databases where we can actually see the data we are storing to the point of being able to index and search it, and even do some filtering operations on it, we have gotten a bit closer to the features of a relational database. With CFDBs we are at a point where we are near relational, but in a very denormalized form.

TABLE 8.9   Hierarchical Document Links

Key (UserId)	Document
10	"firstName": "Alice", "lastName": "Apple", "reporting line": 1
11	"firstName": "Bob", "lastName": "Blundell", "reporting line": 10
12	"firstName": "Charlie", "lastName": "Charter", "reporting line": 10,11
13	"firstName": "Debbie", "lastName": "Drummond", "reporting line": 10,11,12

TABLE 8.10   Normalized Address Table

Id	UserId	Street	City	State	Country	Zip	Type
110	007	123 Main Street	Springfield	NT	USA	80085	Home
111	007	124 Main Street	Springfield	NT	USA	80085	Work
112	007	122 Main Street	Springfield	NT	USA	80085	Billing

TABLE 8.11   Normalized Phone Number Table

Id	UserId	CountryCode	AreaCode	Prefix	Line	Type
21	007	01	636	555	5566	Home
22	007	01	636	555	6655	Work
23	007	01	939	555	8008	Cell

TABLE 8.12   Normalized User Info Table

Id	Alias	First	Last
007	RockyH	Rocky	Heckman

If you look at typical relational database where the data itself are usually separated to the point of being in Boyce-Codd Normal Form [63], which has been a standby normalization form since about 1974. You can denormalize the data a couple of levels, and keep logical groups of it together without having individual rows for each piece of information. Some of the more famous implementations of CFDBs are Cassandra used by Facebook and BigTable used by Google. This is how column-families work.

Using our example from the relational database discussed before, we have this data in normalized tables as shown in Tables 8.10, 8.11, and 8.12.

Relational databases and data normalization are designed to provide very consistent data. They avoid duplicating data, and allow you to only change data in one place. But, in order to find something like the address associated with a person, you have to do a table join. If you have to find linked data across several tables you'll have to have several joins in a query. This can be inefficient compared to just having the data in a document, or in the same table as the target of your query. In column-families we denormalize the table structure and store the related data together in a single column.

Each column represents data that are logically associated such as the parts of an address, or parts of a phone number. Each row is keyed on a value that represents the collection of columns relating to that record. If we were to arrange the data from Table 8.10, Table 8.11, and Table 8.12, we would get something that logically looks like Table 8.13.

To a relational database person, who loves normalization, this is terrible. But, it does allow for very fast entry and retrieval of data, even over millions of rows. The physical structure is more like Tables 8.14, 8.15, and 8.16 which helps this:

TABLE 8.13   Column Family Logical Structure

Row Key	Column Families		
CustomerId	NameInfo	AddressInfo	PhoneInfo
007	NameInfo: Alias RockyH	AddressInfo:Street 123 Main Street	PhoneInfo:Country 01
	NameInfo:First Rocky	AddressInfo:City Springfield	PhoneInfo:Area 636
	NameInfo:Last Heckman	AddressInfo:State NT	PhoneInfo:Prefix 555
		AddressInfo:Country USA	PhoneInfo:Line 5566
		AddressInfo:Zip 80085	PhoneInfo:Type Home
		AddressInfo:Type Home	
			PhoneInfo:Country 01
		AddressInfo:Street 124 Main Street	PhoneInfo:Area 636
		AddressInfo:City Springfield	PhoneInfo:Prefix 555
		AddressInfo:State NT	PhoneInfo:Line 6655
		AddressInfo:Country USA	PhoneInfo:Type Work
		AddressInfo:Zip 80085	
		AddressInfo:Type Work	PhoneInfo:Country 01
			PhoneInfo:Area 636
		AddressInfo:Street 122 Main Street	PhoneInfo:Prefix 555
		AddressInfo:City Springfield	PhoneInfo:Line 8008
		AddressInfo:State NT	PhoneInfo:Type Cell
		AddressInfo:Country USA	
		AddressInfo:Zip 80085	
		AddressInfo:Type Billing	

One of the best things about CFDBs is that they don't all have to have every piece of data. So if your data is missing some bits for some customers and missing other bits for others (called sparsely populated), CFDBs can still deal with it. For example, if the billing address above was missing the Addressinfo:Country that would be fine. Just keep in mind though that CFDBs allow you to create indexes on the column-families. So for example if we created an index on the AddressInfo:Country column in the AddressInfo column family, any entry that was missing the Address-Info:Country value would not be indexed and would not show up in the query that looks for or filters by AddressInfo:Country. This prevents the indexes from storing blanks for any missing information.

Another bit of efficiency in a CFDB is that the information is stored in key order. Your CFDB will order the data according to its key. So when you chose your keys, chose wisely. You'll want to base that decision on how the data are likely to be retrieved and sorted. This will help you quickly find information in the hundreds of millions of rows that CFDBs are designed to handle.

TABLE 8.14   NameInfo Column Family

NameInfo
007:
NameInfo:Alias   RockyH
NameInfo:First   Rocky
NameInfo:Last   Heckman

TABLE 8.15   AddressInfo Column Family

AddressInfo

007:
AddressInfo:Street   123 Main Street
AddressInfo:City   Springfield
AddressInfo:State   NT
AddressInfo:Country   USA
AddressInfo:Zip   80085
AddressInfo:Type   Home

AddressInfo:Street   124 Main Street
AddressInfo:City   Springfield
AddressInfo:State   NT
AddressInfo:Country   USA
AddressInfo:Zip   80085
AddressInfo:Type   Work

AddressInfo:Street   122 Main Street
AddressInfo:City   Springfield
AddressInfo:State   NT
AddressInfo:Country   USA
AddressInfo:Zip   80085
AddressInfo:Type   Billing

TABLE 8.16   PhoneInfo Column Family

PhoneInfo

007:
PhoneInfo:Country 01
PhoneInfo:Area 636
PhoneInfo:Prefix 555
PhoneInfo:Line 5566
PhoneInfo:Type   Home

PhoneInfo:Country 01
PhoneInfo:Area 636
PhoneInfo:Prefix 555
PhoneInfo:Line 6655
PhoneInfo:Type   Work

PhoneInfo:Country 01
PhoneInfo:Area 636
PhoneInfo:Prefix 555
PhoneInfo:Line 8008
PhoneInfo:Type   Cell

CFDBs are really good at dealing with millions of records, or even billions. This is largely due to the scalar nature of the data they store, and not having to deal with joining tables, or performing complex data gymnastics when storing and retrieving data. Sometimes you need to represent some form of nested structure. CFDBs do this by allowing you to nest columns in another column. These are called Super Columns. For clarity; a Column is a piece of data. A Column Family is a collection of like columns stored in a single column. A Super Column is a collection of like column families that store columns of like data.

This break down means that when you update data in a CFDB, only the Column Family is updated. Whereas in a Document Database, you would retrieve the entire document, change a value, and then save the document back to the DB. This makes CFDBs more efficient at updating and changing data than a Document Database.

Because the CFDB data are readable by the CFDB you can query data based on the values of the columns in the column family. This also allows you to do projections in most CFDBs. There isn't a notion of JOINs so you have to request multiple columns from different column families based on the Key.

Continuing our example from above, if we have the following two column families shown in Figure 8.3, the query would look something like:

NameInfo	AddressInfo
007: NameInfo:Alias RockyH NameInfo:First Rocky NameInfo:Last Heckman	007: AddressInfo:Street 123 Main Street AddressInfo:City Springfield AddressInfo:State NT AddressInfo:Country USA AddressInfo:Zip 80085 AddressInfo:Type Home
008: NameInfo:Alias SusieQ NameInfo:First Susie NameInfo:Last Quenterio	008: AddressInfo:Street 148 Balklands Avet AddressInfo:City Springfield AddressInfo:State NT AddressInfo:Country USA AddressInfo:Zip 80085 AddressInfo:Type Home
009: NameInfo:Alias TerryT NameInfo:First Terry NameInfo:Last Tory	009: AddressInfo:Street 1228 Marcott AddressInfo:City Springfield AddressInfo:State NT AddressInfo:Country USA AddressInfo:Zip 80085 AddressInfo:Type Home

Figure 8.3   NameInfo and AddressInfo Families

SELECT NameInfo:Alias, NameInfo:Last, AddressInfo:Street WHERE CustomerId = 008

When you think about how you are going to structure your column families, as always consider how the data will be accessed. The more column families the system has to pull data from, the slower it will be. Therefore you need to organize your column families in a way that logically groups the most commonly retrieved data groups in a column family.

For example, if you tend to work with Customers, Orders, and Hair Dyes most often, you may want to design three column families based on Customers, Orders, and Hair Dyes. You may frequently pull a Customer and his/her Orders. In this case you should consider using a Super Column that contains the Customer column family, and the Orders column family. This will make that common scenario more efficient.

You can create various indexes on the keys, and secondary indexes on the values of columns in your CFDB. This extends to the point of being able to create composite indices that allow you to index combinations of columns such as customer name, order data, and dye color. When you run that query, that combination of parameters will have been indexed and will be able to locate the data very quickly.

To get more efficiency out of your CFDB you can also run Map Reduce jobs on it to extract common data and generate summaries of that data. This can be very valuable in discovering information about the data you hold. Apache Hadoop is one of the most commonly used Map Reduce systems on the market.

Most CFDBs implement their own sharding and partitioning strategies. They typically focus on the key for the column family. In most cases the CFDB orders the information in the DB by key value. This doesn't involve any work on the application side of things. In fact you won't normally see the sharding or partitioning process in action except during setup where you may put different column families on different disks.

As we have seen already, data consistency with CFDBs is eventually consistent. Just like KVDBs and Document DBs, CFDBs don't offer ATOMIC transactions unless you're only updating one value in which case it's kind of a given. You'll need to see if your preferred CFDB provides a mechanism for ATOMIC operations that span multiple columns or column families.

You will also need to consider data versioning to prevent conflicts. Almost all CFDBs use a timestamp on the column family. You can use this to implement the version information for your conflict checking. This will work much as it did when we discussed it earlier in the Service Layer.

8.4.4 Graph Database

Another way to look at how you store data rather than key-value or column families is focusing on the relationships of the data rather than just the data. This is where the Graph databases (GDB) are very useful. A GDB is made up of Nodes and Edges. A Node is an entity like a person, car, or the thing you are storing in the database. An Edge is a relationship between two entities such as family, friend, wife, husband, or in the case of the car example, driven by, made by, garaged in, etc. Facebook is the penultimate example of a Graph Database. It focuses on each person as a node, and the relationships to other people as the edges.

Anything where you want to find a path to another entity, or you want to find relationships between entities is a good case for a GDB. Facebook and LinkedIn use the "You may know" graph traversal to identify people you may want to add as friends or add into your network. Anything you can picture as a tree or a route can probably be built in a GDB. What they aren't good at is calculations of large data sets or analytics based on counts of data. They also work better with pre-defined queries that can be optimized for your particular implementation, rather than random *ad hoc* queries.

Graph databases, because they focus on the relationships, and don't have to perform a lot of data gymnastics discover relationships or perform complex queries are also cheap to run. They tend to be very fast even with a cluster of average computers running the GDB.

This type of NoSQL DB is also more of a logical model than a physical one. Whereas KVDBs, Document DBs, and CFDBs are all dependent on some kind of software running them, a GDB can be built on top of whatever other technology you want to use. You could build one with a combination of KVDB and Files if you wanted to. I suppose, if you were a glutton for punishment, you could even build a GDB with a series of linked lists and file nodes, but you're on your own there.

Each node and each edge can have zero or more properties associated with it to help inform the system of its use. For example, a person node probably has Name, Birthdate, E-mail, etc. A Married to edge may have a "married on" date, and the ownership of the relationship or the direction. Figure 8.4 illustrates this.

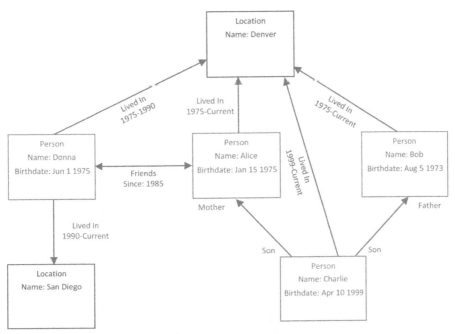

Figure 8.4   Graph Database Sample

You can query this data with fairly complex natural language queries. For example, who lived in Denver in 1992? Who is Charlie's Mother? Who is Bob's Son? It is especially powerful for traversing edges to find relationship depth and connections between entities. The Graph Database is more optimized for pure relationship information than even relational databases. It is not as good as complex data management, and again bulk operations on this kind of data are quite difficult. You essentially have to crawl the graph and perform updates on each matching node.

When you are managing data for your aggregate functions or managing the data as an integrated entity, it's often easier to use a non-normalized structure such as key-value stores, CFDBs, and document databases. They are de-normalized by nature and tend to store related information in a single spot, rather than spread out across multiple tables as a relational database does. This aggregate also forms the basis of your transactional unit.

If you update an entity, you need to ensure that all of the entities the data update succeed. The Aggregate becomes your Atomic, Consistent, Isolated, Durable (ACID) boundary. Similarly that boundary in a Graph Database is where a transaction would span multiple nodes and edges. Operations on a single node or edge tend to be fairly finite and typically don't require transaction semantics.

Indexing in a GDB is usually on the properties in a node. Why not the edges you ask? Well, this is because you can't start a query on an edge, you have to start at one node and traverse the edges to answer your query. So the index is used to quickly identify these starting nodes. Make sure, as always, that you chose indexes and node structures that will satisfy the majority of your queries. Don't do something that seems neat and tidy but isn't what people ask for 95% of the time.

Another quirk with GDBs is that because they are a logical construct rather than a physical one, sharding and partitioning take on a whole new aspect. Normally when you shard a database you break it up on common data structures such as a users' table, a column family, or a key value hash segment. You can do this because the data that you are storing together are all pretty much the same kind of data. But in a GDB, the modes store the data, and they are all pretty similar, but because they are linked to a potentially infinite number of other nodes, breaking them up into shards actually hurts performance rather than helping it.

What you will do instead is group nodes according to how they are accessed or who they are accessed by. You can do this on a regional basis, so that all of the nodes related to the BHD Company employees in Denver are in the Denver server, and the Sacramento ones are in the Sacramento server. While this will improve performance for people in those locations, it doesn't help if you are trying to traverse the graph for the entire organization. Most GDBs don't let you split up nodes like that because it makes traversing an edge that spans regions time out, and that's if it's even reachable by the implementation. So if there is a logical isolation layer, such as one group of node types relates to all things people related in the company, and one relates to all things products in the company, you might be able to separate them along those lines.

Unlike its cousins the Document DB and the CFDB, a GDB is very concerned about referential integrity. In fact, it's the core of how the thing works. If you don't have referential integrity between nodes, then the edges are meaningless. If you delete a node and leave a gap in the neural net of your GDB, it will have seizures and not work very well. Any query that tries to traverse the graph along a path that has a missing node will fail, regardless of which direction it came from. It's like being back in college, and when you had to write that linked list from hand (before they told you about the Standard Template Library), and your list dropped a link without pointing the previous and subsequent links to each other. Suddenly, there was a bunch of stuff you couldn't get to.

To avoid this, GDBs make extensive use of locking, and you need to ensure your changes are atomic transactions. You update nine out of ten nodes, and fail on one, then leave the GDB in an inconsistent state. You really do need to be very pedantic about your create, update, and delete operations. It would be advisable to take a CQRS approach. Have a write copy of the database which then replicates to a secondary read copy in which all reads are performed on. This is especially true if you have a really high volume of data changes. Because you need to lock data and ensure that all writes complete before unlocking it, you will end up with contention bottlenecks. So if you can eliminate the number of things acting on the single set of data, the better the performance.

NoSQL databases are quite versatile. Their value is in their simplicity. If you just want to store data and get it back, a relational database is overkill. If your data update requirements are fairly simple, and you need speed and high-level aggregate storage, NoSQL databases are your best bet. You will probably use a combination of them to achieve a well-oiled system but there is no all or nothing aspect to any of this. You can use any combination of data storage mechanisms to suit your purposes.

8.4.5 How to Choose?

There are no simple answers on how to choose which NoSQL technology you should use. In fact it's pretty likely that unless your system is a fairly simple data-in data-out system, you will use more than one data technology. We'll cover that in a little bit.

As the relational and NoSQL technologies have evolved, they have become more specialized, and some, such as Couchbase have gone from being a very NoSQL key/value pair storage system, to something more closely resembling a document DB graph hybrid. So it can be a bid difficult to pin down which technology to use. What's best for implementing a profile system may not be as good for a status updating system or storing blogs posts or product catalog pages.

There are some high-level guidelines for which NoSQL approach is best for which workloads. Table 8.17 summarizes this.

As you can see, depending on the purpose of a particular portion of your system, you may need one form of data storage or another. There's also a really good chance you'll have a relational database component involved in your system as well. The important part to remember is to pick the right tool for the right job, and you'll have many tools in your toolbox. They can be used together.

8.5 FILE STORAGE

File Storage is exactly what you think of when you think of storing a file on disk. Look at any folder on your hard drive, and that's the stuff we're talking about. One

TABLE 8.17 NoSQL Choices

Technology	Good For	Not Good For
Key/Value	Fast storage and retrieval of chunks of data that may or may not be the same from record to record	Detailed querying, relationships, high numbers of fields. Sorting data by its values
Document DB	Storing and retrieving of serialized whole documents, and being able to query on their contents	Understanding relationships or hierarchical structures documents. Complex nested objects. Bulk updates. Highly transactional systems
Column Family	Storing information with thousands of fields of data and being able to retrieve subsets of those fields based on commonly used queries	Data proliferation, or preventing duplication. Bulk updates are very difficult
Graph	Storing and retrieving data based on the relationships between the various records. Focusing on links between data. Rapid traversal of paths from one data node to another to discover how data are related	Highly transactional systems. Bulk updates. Large document storage. Highly structured data storage

of the best parts about file storage is that it's free. You don't have to buy or obtain a relational database system or even a NoSQL type database system. You just write your information into a file stream and save it to disk. That's a great way to store stuff if all you want to do is store it and read it back again.

Files also have the advantage of being very customizable to your app. Of course that is also a problem because without intimate knowledge of your file format and how you store information, other systems can't read your files. This may be what you want, but in the world of services and making data accessible this is a problem. In an environment where there are lots of standard file types such as .pdf, .docx, .jpeg, and others, flat file storage is very suitable.

If you have basic standard file formats to save, using Binary Large Object (BLOB) storage mechanisms is the preferred method. Yes you can store them directly into a relational database if you want to store the file and associated metadata together. You can designate a column in a table as a BLOB column and submit the encoded binary file date to the DB to be stored in that column.

If the files are small enough, say less than 8K in size, the database will store them inline in its internal column and it will work fine. However if the files are larger than that, or the DBs internal size limit for a column, the DB is going to do exactly what you could have done, it will write the file out to a folder on disk, and store a file reference location in the column instead.

If that is your scenario you may as well write the files out to disk yourself and keep the reference information in a DB table, or even in a NoSQL store for fast lookup. The latter method is preferred in most cases where it's simple file storage such as a collection of images, songs, or some other catalogue type of system. You get to combine the speed of NoSQL storage with the simplicity of file-based storage.

File storage tends to be very fast as long as you have fast disk backing it. If you are saving files to a remote location such as over a WAN, you will naturally get performance issues. There are also things you will want to consider such as file sharing or readability, how much disk space various file types take up and what file types are easier to write and/or read.

Files still take up space like any other data format. When you store things in a relational database for example, you don't tend to think of that. You think of the database getting bigger, but in actuality, its data store file is a file on disk and it takes up disk space. This is why you will want to consider compression as part of your file-based data store.

Compression can be implemented as part of the DAL. However, you will want to implement it selectively. If you are performing compression and decompression on every write and read respectively, you will notice a performance impact. This impact grows as the application traffic grows. Compression should be used for longer-term storage, and for transmission or sharing. The trick is knowing when that is.

Knowing when a file needs to be sent over an LAN, MAN, or WAN is easier than knowing when it needs to be compressed for longer-term storage. You don't want to compress it and decompress it on every read and write. Often, files are compressed when they are closed and their application level locks released. You will also compress files for archiving and backup. Real-time compression is a trade-off game.

If you are compressing files that are used regularly, you will need to balance the performance of the compression, with the efficiency of the compression. Normally compression algorithms that have the best compression, such as 7Z will tend to have the slowest speeds as well. Other systems such as ARC are not symmetrical in their compression and decompression performance. While ARC tends to have amazing compression speeds and the best compression ratios, it is also the slowest at decompression [64]. Great for write and storage, not as great for retrieval and unpacking.

So for "live" compression and decompression, you may want to look at standard ZIP or RAR compression. But for archiving, where there is the expectation of a delay in data retrieval, you might want to consider ARC.

Another consideration is if you bother compressing files at all. There are some file formats such as MP3, JPEG, and most video formats that are already compressed. You won't get a lot of compression out of them for the time it will take you to run them through the compression algorithm. It may just be better to leave them alone and store them. Additionally, there isn't much point in compressing an encrypted file. By its nature, an encrypted file has a high entropy so there aren't many patterns for the encryption algorithms to identify and substitute out.

When it comes to transmitting files to move them, or share them, you will want to ensure that whatever compression you are using, it's a standard that any receiver can deal with. ZIP and RAR tend to fall into this category.

8.6   BLENDED APPROACH

The most likely scenario you will have is a blended approach. You will have some data in files such as images, or zipped archives and you will have some information in a NoSQL store for easy storage and retrieval of document type data, and you will tie it together with relational data stores to provide you any complex relationship management, data normalization, and especially any analytics you want across your data.

For information that will change regularly and contains nested relationships such as a company to its list of products, or a person to their list of addresses, you will use relational databases. They provide a way to prevent data proliferation through normalization, fast look up, and great data management tools.

For the list of products, you may have pre-constructed documents that contain the product information such as name, weigh, dimensions, description, etc. These documents may form a product catalogue. They are ideal candidates for NoSQL data stores such as Mongo and DocumentDb. They can be stored, searched, filtered, and retrieved very easily. You can even do things like get all of the products from all manufacturers that are available for sale in Australia. This collection of documents can form an Australian Product catalogue with one data retrieval query.

Things like the product images, instructional videos on how to use them, or other media files will be stored in flat files in some kind of BLOB or disk-based storage. The NoSQL catalog documents or relational database rows will contain a link to the location of the appropriate file for the related product, company, or reseller, etc.

You may also store or retrieve user profile information from third party services like Facebook. This kind of integration can plug your apps and services into their world in a personal and valuable way. It's not just social integration, you may have third party services and provide part of your solution as well. Inventory services, or pulling stock levels from manufacturer's sites.

Different combinations of storage used together as shown in Figure 8.5 give you a very powerful and very extensible data storage system where you can take

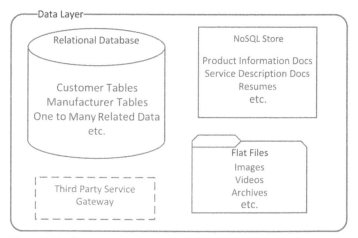

Figure 8.5   Data Layer Components

advantage of the best aspects of each data storage type. Think of each type as a Lego block that you can use to build your data foundation.

8.6.1 The Polyglot Data Layer

When we combine these different data storage types, what we end up with has been called a heterogeneous or polyglot data layer. McMurty and company came up with some good guidelines for deciding which type of data store to use when. As with all things this is not all encompassing, and should be taken with a dose of the Golden Rule. Keeping in mind the comparisons from Table 8.17 let's examine our data layer that we've been using in our previous examples.

In the implementation shown in Figure 8.6, we have a relational database that stores our complex transactional data. It links to serialized XML files, and some binary data held in a file system. The Order table also holds links to the line item keys for our line items that are stored in a KVDB. In some cases these line items contain information on our products which is a view of the catalog page of that product. In our case those catalog pages, being relatively static, are stored in a Document DB as serialized JSON documents. For efficiency we've stored the images from the catalog pages as files in the file system.

We have a CFDB that contains all of the manufacturer information and information about the parts they manufacture. The part page information sheets from the manufacturers are stored in a Document DB like our catalog pages are. We also have a graph database that tracks manufacturer, customer, supplier, and distributer relationships globally.

In this case, we've chosen the best database for the particular workload. Obviously this is a demonstration deployment to illustrate the point of a polyglot data tier. However, it has merit. Each of the storage mechanisms we have here are tailored for

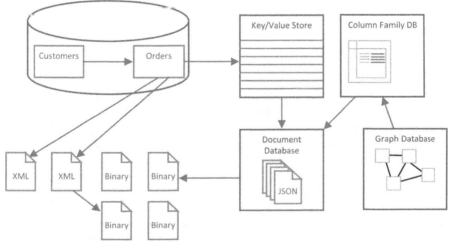

Figure 8.6   Polyglot Data Storage

the type of data they hold, and how that data will be queried. Relational databases are not good at very fast storage and retrieval of documents. Out manufacturers are associated with a reasonably static list of parts they supply and because we do rapid ordering and referrals for parts, it's easiest and fastest to store that in a denormalized CFDB. Because we want to know more about who can deliver what parts to which customer using what suppliers rather than just information about the parts, customers, and suppliers themselves, we have implemented a GDB for tracking and identifying those things for us. We don't need the full information on each of the manufacturer, supplier, customer, and parts in the GDB itself, so the nodes in the GDB refer to the records in the CFDB for that information.

We still use binary file storage for things like images and flat files because it's the cheapest storage solution around and it's reasonably fast. Not to mention that when things like relational databases store files in columns, if the file is over a certain size, say 8K, it actually writes the file to disk and stores a location reference in the column anyway. If we do it directly, we save the extra step at the RDBMS level.

Having a common entry point to a polyglot solution will require a Repository in the DAL to manage it. If you have to call a polyglot data layer from your SL, you'd have to build every kind of data access technology into your services. They'd have to know about KVDBs, Document DBs, CFDBs, and GDBs. They have to know how to structure queries to each type, as well as how to interpret the returned data. That is a tight coupling you really don't want. For any polyglot data layer, you really need to create a solid, versatile DAL.

Your DAL carries the entire responsibility for maintaining the data consistency across the different database types. The databases themselves don't even know about each other much less communicate data consistency rules in a common language. This is perhaps the trickiest part of a polyglot solution. You'll need to pay careful attention to how your DAL manages this issue.

This kind of thing though can be handled through good asynchronous programming, and being comfortable with eventual consistency. NoSQL databases are not designed for Online Transaction Processing (OLTP) workloads with immediate consistency. That process is great for highly integrate data integrity, but not great for speed. When you are dealing with billions of transactions a minute, you need speed and you have to accept that the chaotic dust torrent of incoming data will settle in an orderly manner.

If you need fairly rapid eventual consistency, you'll need to implement the CQRS pattern and have the DAL handle the write side and telling the read side to update when the write is finished. If you use a queuing system for your writes, then the read side notifications will have to be raised by the component processing messages off the queue. This means that the read side will be consistent as soon as it can read the written data.

If you don't need that short of a consistency lag, you can consider database replication and synchronization systems. Most databases of a similar type can be configured to replication between each other, but to replicate information from different types of databases will take a process manager component and code specific to your implementation. Fortunately, this code can be reused, and when combined with metadata mapping, will be able to be converted from one DB pair type to another with replacement of the metadata. This scenario is pretty high latency though. This is the kind of replication you would do from region to region or between systems that do not share transactional responsibilities with each other.

8.7 WRAPPING UP

The data layer needs to be as complex as you need it to be. A large number of very successful systems in use today operate solely on a relational database. Some others like Twitter operate almost exclusively on NoSQL databases. Things like Facebook operate a polyglot solution very efficiently and very effectively. All of them applied the Golden Rule very liberally and came up with solutions tailored to their requirements.

Don't go into this thinking that you need to start where Facebook did. Look at your minimum viable product, and see what will make it the most efficient it can be, without locking your data away so that it can't be used by other systems you may create. Use what you need and discard what you don't. A lot of the data layer implementations are customizable and many of the NoSQL solutions are open source. Don't be afraid to tweak the system to do what you need it to. That's why most large-scale system architects, when discussing their implementations start off with, "We have a highly modified version of blah…"

CHAPTER 9

STRATEGIES FOR ONGOING IMPROVEMENT

IN THIS CHAPTER, we want to take a look at how to keep things rolling once we've built our systems. This is the part that gets overlooked a lot. People often go headlong into designing and developing their systems with no thought to how this thing will be maintained once it becomes wildly successful. Planning for maintaining your masterpiece is as critical as designing it. This chapter is more of a procedural and policy chapter than a purely technical look at the maintenance, but there is some of that too.

I used to be an aircraft mechanic. Part of that job was fixing airplanes when the Federal Aviation Administration (FAA) would issue a thing called an Airworthiness Directive (AD). These ADs essentially said, you must get this preventative maintenance performed on your aircraft or you cannot fly it. One time, one of these came out for a single engine plane (that shall remain nameless). There was a problem with one of the sheet metal structures high in the tail. They were breaking, and planes were falling out of the sky. This is what I refer to as a Committee Generating Event (CGE), and the committee, in this case the FAA, issued an AD to stop this problem.

The manufacturer said that in order to fix this, you had to make a strengthening rib, and rivet it into place inside the tail near the top. But there was a problem. The inspection panels that you could take off and get your hands and tools into, were near the bottom of the vertical stabilizer. In between you and the naughty rib were two other ribs. These ribs had holes in them, but the holes were not aligned. So your arm had to bend in something like an elongated Z just to reach the spot.

This is where failure of maintenance planning made a fixable problem almost impossible and insanely expensive to fix. You do not want to find yourself in that situation. So to save you that, consider some of these issues when you are designing your system.

9.1 FEATURE EXPANSION

One of the most inevitable things you will encounter in your architectural career is adding features to an app or service. It's a death and taxes kind of thing. Fortunately

---

Designing Platform Independent Mobile Apps and Services, First Edition. Rocky Heckman.
© 2016 the IEEE Computer Society, Inc. Published 2016 by John Wiley & Sons, Inc.

this is probably the easiest kind of maintenance to deal with. Adding things to a system is always much easier than removing or changing them. When you add a feature, you just have to make sure you don't break existing features. This is where having a really good set of regression tests is critical to success.

You also need to consider having a production mirror. You want to make sure that your new features won't have any adverse effects on the performance, data consistency, or expected behavior of the existing system. Having a production mirror, even just for testing will allow you to deploy your new features and test them in an exact replica of your production environment. With public cloud providers this scenario is pretty easy to accomplish, and you only pay for the infrastructure while you are using it.

So what should you think about when planning feature expansion? In reality, you need to make sure the feature makes sense in your app. I've seen several people add features to an app more for the sake of getting to learn or play with the technology rather than to help their user base. This is similar to deciding to use a technology because it's new, or because it's the latest buzzword.

I had someone tell me once that they were going to build their company on the Rust programming language (at the time it wasn't even finished). Now I have nothing against Rust, a lot of people like it. They said that they chose Rust not because of how well it suited their solution, their skill set, or that they even knew if it would be finished, but because it was trendy to use it in Silicon Valley. They also said that they weren't using Java, .NET, or PHP because it's too hard to find cheap programmers for those languages.

Let's see here, you're going to bet your company on a language because there are a lot of un-employed programmers that are interested in it? That's because no one is using it for industrial strength applications. There are so many programmers available that are learning it because they don't' have skills that are actually being used in the industry. No, this is not a blanket statement about the capabilities of Silicon Valley developers or people interested in new technology or even the Rust language, it's a statement about choosing the right tool for the right job and making sure you are doing things for the betterment of your product, not because it's trendy. When you add new features, it will take time and effort. Make sure it's something people can get help with, and has a wealth of information and best practices available for it.

You can identify features that your users want to see by using things like User Voice (http://www.uservoice.com). It is used by many software and other product companies to give their users a place to request and vote for features that they want to see in the product. Some organizations have gone so far as to make their User Voice lists, the product feature backlog. Being this accessible, and your users feeling like they shape the product is very good for your PR.

So now that you have decided what new features you want to add, how do you get them deployed? I am going to presume that you know how to develop the features for your app and services. Deploying them so that they are introduced smoothly takes some forward planning. Now, there will be technical nuances to designing your new features. You will need to make adjustments to most of the layers in the architecture for new features depending on the nature of the feature.

9.1.1 User Interface

If your feature requires new UI elements, obviously you'll need an updated app. This updated app will need to be deployed first to internal testers, then beta testers, and finally to production. I presume that you learned about the deployment processes of the various app stores when you initially deployed your apps. Most of them have very good update systems and can notify users when a new version is available.

You will also want to check and see if your new feature took a previously borderline good mobile app citizen, and pushed it over the edge into a battery draining, bandwidth using naughty app. This is where having a really good external beta user base comes in.

You also want to ensure you have a good cross section of each platform represented in this user base. With the proliferation of devices you'll want to test a good selection of them. You can use your user base demographics to find out what the most common devices running your apps are.

9.1.2 Service Interface Layer

If your new features have a new service associated with it the SIL will need to expose that service. On most cloud platforms you can add services to your SIL without having to disrupt communication from existing clients. If the service is a new method on an existing service, it likely means a service redeployment and a restart of that node. If you do have to go through an update that requires any node restarts, use the cascading update pattern discussed previously.

9.1.3 Service Layer

Your SL is where you've deployed your new or modified service. You will probably need a cascading update on this layer as well depending on if you are adding new services or features to an existing node. If you are adding new nodes entirely, you'll be able to deploy them and just tell the SIL that they are ready to accept traffic.

9.1.4 Data Abstraction Layer

The DAL work would have been done prior to service deployment into testing and production. Depending on how you have built your DAL, some of the physical components may be deployed in the SL, and some in the DL. Or, if you have made a clean separation, it will have its own infrastructure layer.

While the third option is best for maintaining that clean decoupling, it does mean a bit more infrastructure management. If your system is fairly small, you will likely co-deploy these components with the SL and DL. At such time as you are running a large system that needs 24/7 uptime, with significant traffic and transactions per minute, you'll want the DAL to be a separate layer that can be updated independently.

This will help isolate changes and give you more freedom to perform the updates, and it means you can shuffle layers around to get the best performance and protection of your data. This is eventually where you want to be, but it may take some growth to get to the point that the return is worth it.

9.1.5 Data Layer

The DL can be tricky to update. In relational databases, schema additions aren't a big deal and can be done with little or no downtime, but changes to existing structures or columns can wreak havoc if you don't have a good DAL acting on behalf of the calling services. Getting the DL and DAL coordinated is super critical to ensuring a smooth upgrade deployment.

NoSQL upgrade complexity varies from type to type. For Key/Value pairs it's almost a non-issue. You are probably just adding a new KVDB, or adding extra information into the values being stored. Even if you are changing the information in the values, the Key-Value Database (KVDB) itself doesn't really care. If, however, you decide to change your keying information, then you will have issues. You'll need to migrate the KVDB from one instance to another, storing the old data in the new instance with new keys.

Document Databases are similar, in that adding information to the documents is not a big deal. You will have to add new indexes, your new information is a primary query target though. Again, if you change your keying structure, you'll have more to contend with. So be careful with that. It's a really good idea for you to carefully plan out your keying approach on any database so that it is as resilient to change as possible. If things change too dramatically you might want to consider just creating an additional DB to hold the new versions and once all of your users have migrated to the new system, kill off the old ones when you are able to while meeting your archival compliance regulations.

Column Family Databases (CFDBs) are a bit more complex. Changes to the column family structures are considered a significant update. Some of the CFDB systems actually require you to tear down and rebuild the column families if you change something. This again is a case where you'd want to stand up a new DB, migrate the data into the new structure, and turn off the old one.

Lastly, Graph Databases are pretty resilient to additions. If you add a new node type, it's not like there's a bunch of existing nodes you need to touch to update. It's the same with edges. But if you change the properties of a node, all of those nodes need to be updated. A new field in a node will result in previous nodes not having that information. This means a graph traversal that will return that information needs to be protected against blank data.

If you have a query that starts on one of the new node types, and relies on that new information, then query isn't going to get very far if it starts on an old node type. Keep in mind that a GDB is a logical construct and what is required if you update the structure is largely dependent on the underlying technology that you've built it on.

9.2 DATA COLLECTION MATTERS

Obviously the first thing you want to do is develop, test, and deploy the service side into a production mirror. Then you want to test it some more. Once the unit tests and headless test have passed QA, you'll want to introduce the feature into your updated client. Then you'll perform copious amounts of internet client testing. Once that passes, you'll let some of your die hard early adopters beta test it. At this point you want to collect as much metadata and performance data as possible about your system.

The more performance data you can collect, the better you'll be able to predict performance bottlenecks at scale. You also want to have that subset of your users, call them your insiders, which can help you test the product in real world scenarios. Most app marketplaces have a beta section where you can give specific people invites, or direct links to your app so they can download and test it without the rest of the world seeing it's there. Take advantage of that.

I have been asked why I don't recommend just dropping two versions of the system out there and AB testing it with the general user base. Mostly the reason is because in this context we are talking about apps, which have to be deployed to devices. If we were working with a website, then this would be a really good way to test feature acceptance, but even then, it's not great at testing performance or system integration. With mobile apps, it's much harder to get good AB test because you have to redeploy the app each time you switch things up.

This is also the reason that you want the service side production ready by the time you start deploying test apps. If it's not, and you have to keep fixing it, you may have to update the app to accommodate the service fix if it was something to do with a data structure, or how the app called a service. This round tripping gets very expensive in time and effort wise.

Once you have the app in the bands of your beta testers and you start doing performance testing, you'll know what kind of scale adjustments you need to make on the service side. If the feature is simple it may not require you to change your scale patterns. However, if you have added a feature that turns a 9 to 5 app into a 24/7 app you may need to adjust your service scale to be at a higher level, 24 hours a day. This is also true of your support structure.

If you provide support for your apps and services and you have been running a 9 to 5 call center, you may need to change that to around the clock support. Of course if your app is a game, or something not mission critical this won't apply, but if your app support say, ambulance services, it will. So consider the ramifications this will have on your non-IT infrastructure as well.

Once the beta testing is stable, and you have gotten you bug count down to a reasonable level, you can deploy the app to a larger testing audience, or if you are confident, deploy it into the production marketplaces for the various platforms. At this point, you aren't quite done. It seems that there is always that one user, or that one bug that never had a problem, until you've decided it's your golden deployment. So make sure that the excellent and robust bug reporting system you've used from the beginning with your app is equipped to deal with the new feature.

It's a simple thing, but very often overlooked. Sometimes a new feature gets deployed into a product, and the support staff get a call about a bug in the new feature. But they can't accurately report on it because the feature isn't listed in the bug tracking software, or they haven't been trained in dealing with it. Don't overlook this.

If I were to break this down into a list, it would essentially be these nine steps.

1. Identify viable worthy features
2. Design the app and services to support the feature
3. Build and test the service in a production mirror
4. Build and test the enhanced app
5. Deploy the app to beta testing group
6. Performance test the system as much as possible
7. Adjust service layer for scale
8. Deploy the service into production
9. Package and release updated app with new feature

If you add a new feature that just adds one extra column to a table in a database, the schema update will be easy, and you won't break old clients because they'll ignore the extra data. But if you have to change the schema, or add a new column family, there will be downtime associated with the service upgrade.

I talked about this cascading upgrade in Section 6.9.2.3. Any time a service has to be taken down to upgrade it, you want to use a cascading update pattern to perform the service updates. You can re-read that section if you skipped it, but essentially you have a load balanced set, you migrate traffic off of one node, upgrade it, reboot it, and add it back into the rotation then rinse and repeat for the other nodes in the cluster.

9.3 MULTI-VERSIONING

Now that you have decided to upgrade your system. You are going to have to consider how you go from version to version. There's always the simple way, you deploy the new one, and then tell everyone to upgrade in 2 weeks because you're shutting off the old one, but that will be a user abandonment event unless your service is something they don't think they can live without.

In most cases you will have to run multiple versions of your mobile apps and services even if just for a little while. Being able to run different version side by side can be costly when it comes to having infrastructure to support multiple versions. There are also the cases where you want to have a tiered set of offerings for different levels of customers. In these cases, you are not so much running different versions, but running different entitlements. Let's discuss this first since it's one of the more common questions I get asked.

When talking with ISVs that are moving, changing their product from an on premises deployment to more of a Software As A Service model, including creating new mobile apps as the new UI layer, they aren't sure how to go about making that change. Initially, they all think they'll just take their on premises deployment stack and stick it in some virtual machines in the cloud.

This VM approach works, kind of. When they try it, they realize that there are issues with user authentication to a local authentication store, or that they need access to files on the customers' SAN. Or perhaps their software that works great on a LAN isn't so good over a WAN. The biggest problem though is going from a clear single tenant installation, to wanting to take advantage of a common app and service layer for all of their SaaS customers to a back-end database and how to make that single database multi-tenant.

In the beginning, your only quick win will be to stand up an entire stack for each customer, including their own database. Usually, because customers want some assurance that their software stack is isolated from everyone else, it means they get their own database server as well. So step one is to migrate the existing deployment into a cloud hoster as-is.

Once you have that up and running and you get a few customers on it you can think about how to start breaking this up and running it with a common UI and Services layer, and still give each customer their own database. This results in you running two versions of the service.

Your tier 1 version is a full dedicated stack per customer. Your tier two version is a common UI, SIL, SL, and DAL layer, with each customer getting their own DL. The data can still be put into individual database servers (hosting relational or NoSQL data stores) but the code layers are shared. Making this step actually helps you learn to become a services company instead of just a software vendor.

The third step is to have a fully multi-tenant solution. All of the layers are shared. While you can still offer individual databases to each customer, they are on a common server cluster. This is a full SaaS offering in a multi-tenant configuration. This is the most economical way to run SaaS systems. However, it is also the one that is hardest to update and deploy new features to, due to the large customer base on the single implementation. If you add a service, every customer gets that service.

What you end up with is three versions of your product running side by side. By being able to offer that 'shared everything' implementation, you will be able to offer your SaaS service at a much cheaper price to a whole new customer base that probably couldn't afford your software if they would have had to buy servers to install it on premises. Your mid-range customer base still gets the advantages of a hosted service provided by you, so they no longer have to host their own infrastructure, but because they can pay a bit more, they get their own data tier to isolate their data. Then your high end customer get dedicated everything. You can largely just pass on the extra infrastructure costs to them as part of your hosted service. They'll likely be fine with that because they don't have to maintain infrastructure of their own to support your product on premises anymore.

These three versions are relatively easy to maintain as separate versions because the delineation is very clear and changes to one version won't affect the other versions. You can add a shiny new Data Visualization system to your top end dedicated product and neither of the other two versions won't know if even happened.

This is also a case where the lower tier shared everything will really only be running one version at a time. All of the customers operate on the single shared deployment. When you upgrade it, everyone gets the upgrade. There is no notion of multi-versioning within this tier.

The mid-range version may run multiple versions side by side. However, this should be a rare thing, and the complexity is in the shared code. The individual data layers are fine with multiple versions because they are on a per-customer basis and not shared. The UI, SIL, SL, and DAL though are shared. This will mean running a set of this shared infrastructure for each version you want to maintain. You will have one UI, SIL, SL, and DAL deployment for each version you have in operation. Due to the larger scale of these implementations, this can be costly. I would advise you to keep your versioning strategy to an N-1 or N-2 at most. Running more than three versions of your shared infrastructure would not just be prohibitively expense, but also difficult to support. This is magnified if your low-tier and mid-tier versions are different from your top-tier single tenant versions as well.

In the top tier, the customers will be paying top dollar for your services. So if they want to hang on to an older version, you can reflect this cost in their bill. Almost every major software vendor does this, it's called Extended Support. Microsoft is the perfect example of this. They have an N-2 policy. The current version and the previous two versions are actively supported. Beyond that, you have to purchase extended support. Anything beyond the extended support range and you are on an individual contract basis. Those are very expensive. Just ask anyone who still had a Windows XP support agreement in 2014.

The other scenario we want to consider is the consumer facing scenario. If you are writing apps and services that are sold through the app stores directly to consumers, you are in a different operating model. In this world, you don't get old versions of apps. By the same token, you can hang on to an old version of an app for as long as you like. To put that in perspective, every single individual has the option to never update their apps. This means that you could potentially have users running every single version of your app that you ever released.

This sounds daunting, but the reality is actually pretty simple. For anything other than major apps or mission critical ones such as Office Productivity, most app developers simply change the system, release the new app, and turn off the old one. When users complain their apps don't work anymore, you tell them to click "update" and get the new version.

This is actually the promise of the new app economy. You buy an app once, and you never have to buy it again. You just keep updating the app. As new features are released, the updates just keep coming and keep making the app better. Even major software vendors such as Adobe and Microsoft have gone this route with their flagship products. You don't buy them outright anymore, you pay a subscription fee, and the products are continually updated on your machine as long as your subscription is active. So multi-versioning of apps themselves can take care of itself if you buy in to the new world order, which, I recommend you do.

There are still reasons to multi-version though. One is money. You can release a version of your app and some updates, but you may decide to release a new version with pretty major upgrades. At this point you might consider it worth having people buy the app again. Be warned though that the precedent in the industry for this is pretty rare. If you do want to do this to help gain another revenue spike, the update will have to be significant. You might only do this if the changes were dramatic enough to require you to set up an additional set of infrastructure. At which point you'd have to almost be able to present this as a new bit of software. Most developers that do this often have a free version and the "Pro" version of their apps. Then if they do it again, they have an Ultimate or Enterprise version.

In any case, your SIL will be responsible for routing the right app to the right version of the service. This in itself isn't too difficult. You're going to need a versioning strategy for your services. As we talked about when we were discussing the SL with RESTful web services, there are different approaches to how you version them. As I said before it depends on if you are versioning the resource or the service. You can do resource versioning with Accept headers and the like.

Service versioning is a matter of choice. While there is no hard and fast rule to how to version service endpoints, there seems to be an emerging trend to use api/ver/service type URLs. For example, http://www.bluehairdyecompany. com/api/v1.0/order or http://www.bluehairdyecompany. com/api/v2.0/order.

Your service versioning can easily follow a pattern like this to provide separate self-describing endpoints. The apps themselves can use configuration information to stick to the version they were designed for, and when the app is updated, the service version information in the configuration can be updated as well. It's important to keep in mind that changes to your apps and services need to be versioned in a clear and ideally common sense manner.

9.4 VERSION RETIREMENT

Eventually, your client apps will be updated and your users will have migrated to the newer versions of your apps. With any luck your upgrade strategy will lure stragglers so that you don't have many left behind. If you do end up in this enviable situation, you can just turn off and undeploy the old services. The apps pretty much take care of themselves. The idea with most app stores is that when you release a new version, the old one is no longer available, unless you have released different versions as different apps, like Lite and Pro versions.

However, if you are developing apps for a company or corporate customers, you can't just turn things off because it's quite likely that there is some customers out there that have a custom app written against your services, or they never purchased the subscription so aren't on the upgrade path. In the case of the latter, they really can't complain if you disable an old service that they aren't paying their subscription for. Don't worry though, they will.

As I mentioned above, you need to have a vision as to how many old versions you will have the capacity, or interest, to support. If you have a successful app, and you have many long years of success and new versions, at some point you will have more historical versions that you have resources to support them. So you need a version retirement plan and your users or customers need to be aware of it.

When you start to plan a feature upgrade that will warrant a new major version, you need to include user notification of your N-2 (or whatever number you chose) policy and that the release of the next version will make the version that is currently N-2 fall off the support truck. Although you probably want to word it differently.

As part of your upgrade documentation, you should include documentation that explains how to upgrade from version N-2 to the current one. Over the years I've always found that there are essentially three upgrade philosophies customers use. Well four if you count not upgrading at all.

1. The first one is always staying on the latest version. These customers are usually pretty technically savvy and are probably going to want to be in your beta testing programs as well. If they have longer upgrade cycles they tend to upgrade to the latest version on the market at the time they upgrade.

2. The second one is a bit more purposeful with longer upgrade cycles. Essentially, you install the latest version at the time you are ready to do the install/upgrade. Then you tend to sit on that version until you are basically forced to upgrade again due to support or technical issues. When you upgrade to the latest version, despite the teething problems, it usually means you have a long span of stability before you have to upgrade again.

3. The third one is the typical "We never take a .0 release of anything." This is the ultra-conservative version. This is how many government departments operate. They can't afford instability so they usually wait for the first point release that includes the hot fixes and major bug fixes. But this also means that they lose some of the length between upgrades because they tend to ride closer to the end of the support cycle. These are typically the ones that will pay for extended support, even if it costs them more than the internal and external issues do with .0 releases.

You'll have to line up your retirement policies with the types of users you have. If the majority of your users are type one, then your older versions can probably be retired by just switching off the services because your customers will already be on the latest release. You can then retire the infrastructure and save some money.

If you have mostly type two customers, then you're going to have to deal with having multiple users on multiple versions. This is not ideal, but the number of versions you have to support is fewer than the list for type three. With type two customers, they tend to leapfrog the versions, so you'll need pretty good support for big bang upgrades. Customers are likely to go from version 1.x to version 3.0. This can be tricky if your apps and services assume that the upgrade will be from version 2.x. So when you retire version 1.x, you'll need to have tools, or scripts available to migrate customers from 1.x to 3.0. You can do this either by having them go through an intermediate "install 2.x" prior to installing 3.0, or preferably, ensure the system upgrade tools can migrate directly to version 3.0.

For type three customers, you really need to have a great support program. Not only because it will be used a lot, but because larger customers are likely to want extended support once they get comfortable on a version that they use or even modify for their use. You should plan for the extended support scenario, and make your upgrade path very attractive. If you get a reputation for really stable .0 releases your customers may upgrade on a faster cadence, but don't count on it.

When it comes to actually decommissioning the services themselves there are a few things you need to do for your customers to make the upgrade smooth and entice them to upgrade. These mostly center on the upgrade process support and tooling. In this context we are mostly talking about commercial and in-house apps and services. If you are targeting the general consumer market with apps which are the only things that talk to your services, you pretty much just deploy the new app version and move on, but we'll cover that in a bit.

If your upgrade requires changes to the UI layer, you need to have some kind of in-built training or walkthroughs of the new features. If it's a pretty major release and your customers are large enough you'll want to have either self-paced online (video, or presentation capture) or in-person training available. This will give the users a sense that you are more than just your app, and that they aren't left in the dark. Microsoft's Windows 8 demonstrated what happens when you drastically change a UI, and then abandon the users to figure it out on their own.

You also need to have good training and tools for the IT staff if there are any business to business systems interacting directly with your services. If you have created your system as a SaaS service layer, where the users don't have any of the data locally, then the data migration is entirely up to you and you've probably already done that as part of your new version development. But if the customer keeps any data, or they were allowed to install your system on-premises, you'll need to provide detailed support and ideally migration scripts that will update the data layer and perform any required data migrations.

You can leave the older versions of the services running until you reach a break-point where there aren't enough users actively using them, that you can just shut them off. That threshold is up to you and your business requirements. For some organizations it's down around zero active users. For others like Google, they e-mail people telling them that the service is being shut off on X date, then they shut things off on that date regardless of the number of active users. In a free system this kind of works because you aren't losing revenue by cutting those users off from the service. If customers are paying for a service though, you need to warn them long in advance, and send them several notices that include instructions on how to prepare for the upgrade, and where to get help.

9.4.1 Scale Back

One of the ways that online games tend to retire versions, or entire games for that matter, is to scale back the servers until there are only one or two running to handle the stragglers. This is actually a good strategy if you have customer types two and three. Your service cluster numbers will be a bit of a bell curve to fit your particular adoption cycle.

For a mostly type one customer audience you'll want your version infrastructure distribution to look something like Figure 9.1.

You will have more horse power running your latest version. Most of your customers will be moving to this version and chances are they'll all move near release time. So it's feasible that you'll also have a pretty big initial spike when you release. You may need additional on-demand capacity for the first month or two after your release until the load moves back into a business as usual range.

For type two customers, the cluster distribution changes a bit of a flatter distribution as in Figure 9.2 because unless all of your customers signed up and started their upgrade cycles on the same day, at any given point, someone will be ready to upgrade.

Type three customers are going to have more of a sine wave distribution like that in Figure 9.3. They'll get onto a point release, and sit there until they really have

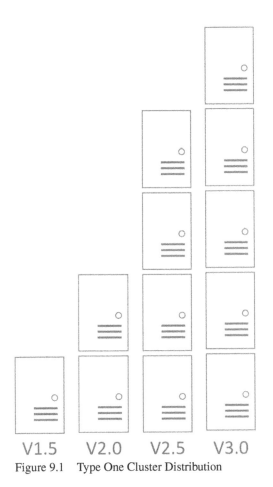

Figure 9.1   Type One Cluster Distribution

to upgrade. Keep in mind that they are likely to want extended support as well so you may very well end up with more than N-3 versions.

While you are probably going to have auto-scaling turned on for your services, you're going to want to limit the scaling to something that is expected rather than have capacity and disk space waiting to be used. I wouldn't suggest that you allocate smaller VM sizes to the older versions and limit their performance, as an incentive for your users to upgrade. This is somewhat unethical and is likely to lead to customers going somewhere else rather than upgrading.

You want to lead people with a carrot, not drive them with a stick. Entice them with the new features. Make sure that your new versions are enough of an improve-ment to where they really see the value in upgrading. Then you can retire your older versions much faster, and it will make you more efficient.

Depending on your margins and if you can keep the infrastructure running with-out eating too much profit margin, you will want to scale back servers on older ver-sions as much as you can without affecting performance. In some cases, you will want to officially "retire" a version, but perhaps leave one server running with that version

V1.5　V2.0　V2.5　V3.0

Figure 9.2　Type Two Cluster Distribution

for a limited fixed period of time. If you are very customer focused, you can identify who is still using that server and contact them letting them know that it is being or has been retired and will be shut off in the near future.

9.5 CLIENT UPGRADES

A quick word here about upgrading the client apps. In most cases if you are deploying your apps through a platform app store, this involves packaging the updated code, signing it, and uploading it. There are some nuances though which of course make the process slightly different for each platform. A few of these are listed here in Table 9.1.

The slight differences aren't dramatic, but they are enough to require attention. It's interesting that Google Play requires you to sign the app manually, and doesn't require any app certification. You also have to name the package exactly the same which is how Google manages apps and you upload it directly to the same record in the Google Play Developer Console rather than creating a new version record for the app. Both Apple and Microsoft automatically sign the package with your developer credentials when you build the package, but require an extra certification step. By submitting apps into your developer record on the store, and updating that record it doesn't matter what the file is called, it will still be listed as an updated version correctly.

You might also want to note that for Apple, they will list the current version and one previous version of the app in the App Store. Microsoft and Google don't do this. All the stores allow automatic updates to new versions if the user has enabled that feature. This can be tricky with business-related apps on controlled devices.

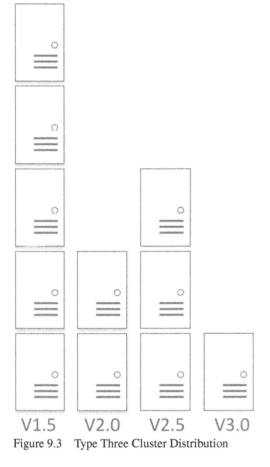

V1.5   V2.0   V2.5   V3.0

Figure 9.3   Type Three Cluster Distribution

TABLE 9.1   App Store Update Nuances

Apple App Store [65]	Google Play Store [66]	Windows Store [67]
Update your version information and repackage the app.	Create your new app package. The APK Package Name must match the current version.	Update the version information in Visual Studio and repackage the app.
Create a Next App Version in iTunes Connect.	The new Version Code needs to be greater than that current version.	In your Windows Store account select the app and select Create New Release.
Upload the binaries for certification.	You need to digitally sign the updated APK with the same certificate as the current version.	Upload the new package into the app record, then select Submit for Certification.

In some cases, the corporate IT policy will disable automatic updates of apps for security and consistency reasons. For this reason, if you release updated versions of the apps you may need to send notifications to corporate users who are using your apps from the app store. In cases where you side-load apps, this isn't a big deal because corporate IT will be involved in pushing the apps out to the devices anyway.

Some things you will want to consider when you upgrade your apps to ensure not only that they are downloaded more, but that you keep happy customers.

Do your homework. Check your ratings and user feedback. Make sure that if there is a trend toward low one star ratings because one of your apps' core features is broken, that you know this and address it. See if there are common requests, or if a particular user demographic has an issue with your app.

Monitor ratings and download frequencies. There is a relationship between downloads and ratings. Bad ratings can affect downloads but it starts before that. High downloads and poor ratings mean broken functionality or misleading app descriptions. Low downloads and high ratings mean poor promotion and visibility. High downloads and high ratings mean money in your pocket. Try to get to that one.

If things are slow for your apps, you might need to create a "bait app." A bait app is one that exposes your users to some of the cool things your app can do for free. There are a couple of versions of this. You can provide a limited set of sample data with the app to demonstrate all of the functionality, but with data that are not pertinent to the user. Alternatively you can provide limited duration or limited functionality but with their data so they can get a feel for how it would work in their lives. This second version is pretty good if you play it right. Limited duration apps are just frustrating. Once they expire, the trend is for people to just delete the app. With limited functionality, you appear more benevolent and the user gets hooked on your base features, but the one feature that they really want is in the paid version. The conversion rate on these kinds of apps tends to be higher.

For example, Game conversion rates will average between about 1% and 30% [68] in 2011. This has changed over the years when people started to expect free apps that were actually worth downloading. The new app economy has changed everyone's perception of value for money. For example, Zynga a poplar mobile app game developer averages about 1.7–2.1% [69] conversion rate from free to paid users. While this doesn't seem like much, when you consider that there are around 130 million users of Zynga's games and they convert about 1.7 million of them to paid, it will certainly keep the lights on.

Converting users is much easier if things are more attractive or they have a reason to upgrade. Better features, are one, but additional features like reports or integration with other systems tends to make the upgrade not only more attractive, but easier to accept from a payment standpoint. Some people will release a free fully functional version of their app with limited data, but then charge a subscription fee for their back-end services that connects the app to real data. The best version of this connects it to systems that corporations would want to purchase subscriptions to such as HR management, financial management, inventory management, etc.

Now that you know what people want in a new version, the other thing you need to do is fix your bugs. Yes, sometimes this actually needs to be said. Fixing your app bugs actually starts with the previous release. Not necessarily because that's where you introduced the bugs, but because that is where you should have introduced good crash reporting. If your users can submit crash reports to you, fixing your bugs become a whole lot easier. Some of the app store submissions allow you to upload the symbol files with your app, yes this makes the package a bit larger, but the data you get out of it makes your maintenance and updating so much easier, it's worth it in all but the trivial cases.

Once you have your bugs fixed, match that up with the demographic and ratings data you collected before. Prioritize the ones that are causing the most user pain, or anything related to security. Once you've addressed them, make sure you highlight them in your release notes when you upload your new version. Make sure people know you've been listening, and that you care about creating a good product.

If your apps are growing and you think you should expand, adding additional regional capabilities can be a good step into expanding your audience. This can be a bit daunting though, especially to do correctly. One of the problems I see with a lot of apps is that when they decide to go global, they create a version with a new language in it. Instead of getting someone that actually knows the apps native language (UI language that is not C#), and they know the target language fluently, they use something like Google Translate to convert their text into the target language. While these kinds of things can translate words, they aren't always as good with grammar. This is how we ended up with things like "Chinglish" [70] becoming a meme.

If you do this, try to get an actual linguist to help you if you can afford it. If not, at least get a native target language speaker, who can communicate with you, to do the translation. Try to avoid getting someone who took two semesters of the target language in high school to do your translations. They might have forgotten a few things.

Last but not the least, in fact probably one of the most significant things you can do for your apps is to remember that looks matter. Apps that have really good sample graphics, good logos, and attractive UI screen shots get downloaded more than those that don't. Choose screen shots that show off your apps sexiest visual features. Make your app icons elegant and attractive. Don't skimp on the visual appeal of the app. By the same token, you should get professional help with this if you can. If you can't, don't go beyond your graphics capability. Simple clean elegant UI is still better than a gaudy 1970's acid influenced color scheme. A friend of mine who is a graphic designer told me a good rule of thumb once. You should rate yourself as a graphic designer on a scale from 1 to 10 with 1 being the lowest. Then never use more colors than your UI designer rating.

If you suck at graphics design and rate yourself a 1, and you need more than one color in your app, phone a friend who knows graphic design or look one up on fiverr (https://www.fiverr.com/) or 48hourslogo (http://www.48hourslogo.com/) or something like that. Even sites like freelancer (https://www.freelancer.com/) can connect you with talented people who will work on a small contract basis.

A good app maintenance strategy will go a long way toward making sure you are successful in the future. Apps don't often get more than once change to make a good impression. This means that you obviously want to focus on your initial release, but if it's good enough, users will give you a chance to fix anything that's broken, or add new exciting features as things come up. Updating your apps and showing your users that you are listening to your feedback will provide some insurance for a continued growing user base.

9.6 WRAPPING UP

There aren't many people that go into creating an app and services, just to launch it and walk away. Overtime your needs will grow, things will change, your system will evolve. You need to plan for that expansion. That planning starts with designing systems that can be maintained.

You do this through good telemetry on your applications, and being able to connect with your users to find out what they really want to see out of your apps. You can triage your feature expansion through user feedback and data collection. Once you know what you need to do, you can expand your features and be able to deploy better products.

CHAPTER *10*

CONCLUSION

$\mathbf{W}$E COVERED A broad spectrum of things to consider when building future proof apps and services. In today's world where platform leaders will change yearly if not more frequently you have to be abstracted from platform specifics as much as possible with your apps and services. If you aren't, you won't be able to keep up with the changes and you'll fall farther and farther behind the competition. This is a very fast paced industry, and agility is a core skill for any organization if they want to survive.

There have been many attempts at cross-platform technologies and approaches in the past. Some of them like Java did pretty well, but still aren't agile enough to survive changing platforms with a significant lag and fragmentation of the runtime. Plus, focusing on one lower-level technology like this can lock you out of other platforms such as iOS and Windows. We need to think at a higher level to avoid painting ourselves into a corner.

This higher level starts at the architecture and technology choices you make before you start writing any code. You have to think platform agnostic from the very beginning. This approach means you have to consider the architecture patterns and tools you need to achieve platform independence. You also have to think very standards based.

If you make sure your design decisions take a standards first approach. Use RESTful web services with standards like JSON, and OAuth. Once you align your thinking to this, choosing technologies is easier because you have a list of standards that it must support. Your cloud services platform should be as standards based as possible to avoid too much vendor proprietary code. This will help you avoid some level of lock-in.

That being said, you can't avoid it all. You need a balance between capability and lock-in. the more of a single vendors stack you can use, the better the integration will be, and the more efficient your systems will be. This will help inform your decisions on vendors for your cloud services. If they are standards based as much as possible, while offering a good integrated end to end stack, then you will have the flexibility to be as future proof as technology can be.

With technologies and vendors selected you can think about your cross-platform development tooling. There are several cross-platform development tools available now. Just about all of them can cover Android, iOS, and Windows app development. Tools such as PhoneGap, Xamarin, and Visual Studio all support cross-platform app dev very well. You will still need something like Visual Studio or Eclipse

---

Designing Platform Independent Mobile Apps and Services, First Edition. Rocky Heckman.
© 2016 the IEEE Computer Society, Inc. Published 2016 by John Wiley & Sons, Inc.

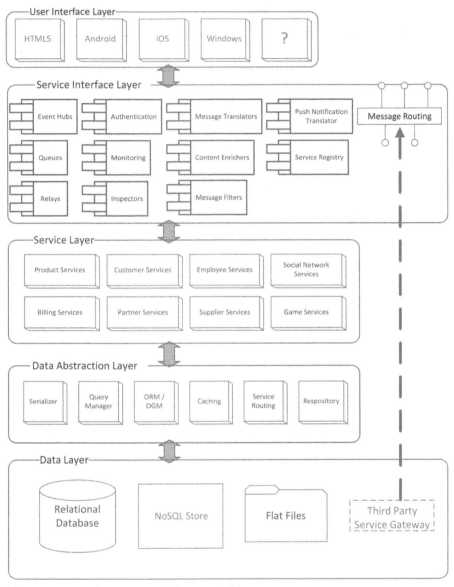

Figure 10.1   The Five-Layer Mobile App Architecture

to do your back-end service development, but since its server side, it can really run on whatever you are familiar with. Make sure you can use these tools to use SDKs for things like mobile application services, push notifications, NoSQL data store access, and of course standard RESTful and SAOP/XML web service creation.

You can build out your apps and services with the five-layer architecture approach to give you the best possible chance at future success. Figure 10.1 shows a complete stack for the best agility and future proofing of mobile apps and services.

The additional Service Interface Layer and Data Abstraction Layer in addition to the traditional Presentation/UI, Business Logic/Service, and Data layers will provide the right levels of abstraction to allow you to adapt to whatever platforms, client or server, that come up in the future. This division helps you deploy any or all of your apps and service across a combination of device, cloud, and on premises component deployments. This gives you the broadest range of supportable scenarios from public-facing websites and consumer apps, to commercial apps and business to business services, to government and regulatory controlled apps and services where the data layer has to be kept in the customers' data center.

The abilities of the SIL and the DAL provide your apps and service with the ability to not only connect to whatever device the world can come up with, but to expand your business and business model. You can offer full apps and services direct to customers, or allow business partners to connect to your services directly. This adds a new dimension to your business capability. You can offer other partners the opportunities to create their own app UIs and connect to your services directly as a value-added app vendor. You can partner with other organizations to offer blended data and information for big data systems and research.

You can easily create different levels of service offerings for different market tiers. If traditionally you only served mid-market customers because they were the only ones that could afford servers to host your apps and services, now you can expand that by hosting your services in the cloud, and offering the app tier for installation on commodity mobile devices. You can reach a much broader user base since they no longer have to purchase infrastructure to host your systems.

You can expand into higher tier markets because you can access the internet-scale power of cloud computing for your service back-end that larger enterprise workloads will require. With cross-platform app development you can deploy either through the app stores, or directly to the devices through corporate IT to ensure the best level of customer service and broadest reach for your apps.

You can also use this SIL and DAL abstraction to connect to other external services yourself. You can now pull in whatever information and services you can reach on the internet to enhance your app experience for your users. You can create customized data flows from a combination of data services, and serve them to apps in new and meaningful ways.

You can tap into Facebook, Twitter, Instagram, and other social network providers to create agile future proof apps and services that become part of your users' lifestyle and integrated into their lives. With cross-platform development tools like PhoneGap, Xamarin and Visual Studio, you concrete these apps that look and feel native. You can integrate with devices so that you are enhancing the users experience on whatever device they have at the moment.

By concentrating on your back-end services, and using a minimalist approach to the amount of code you deploy in the device apps, it also means your users' experience is portable from device to device and platform to platform. With your apps, they no longer have to leave their data behind when they switch from one platform to another as they inevitably will. If they first download your app on an iPhone, and then change to an Android when their contract rolls over, if they know your app is available on every platform, and that their data will transfer from one to the next, they will use your app instead of someone else's. Think of things like Dropbox and OneDrive. It doesn't matter which device you use, your data go with you.

We are in a new world of mobile computing. Apps are the new software and services are the new IT economy. If you aren't building your development plans on platform-independent delivery of cloud-based services, you will be an ex-company in 3–4 years. Plan for the future, build apps for every device, from a single codebase with powerful cloud-based services, and you will succeed.

REFERENCES

1. M. Lacey, "Windows 10 Developer Announcements from Mobile World Congress (From Someone Who Was There)," Microsoft, March 5, 2015. [Online]. Available: http://www.microsoft.com/en-gb/developers/articles/week01mar15/windows-10-developer-announcements-from-mobile-world-congress [Accessed April 15, 2015].
2. J. R. Robert van der Meulen, "Gartner Says Global Devices Shipments to Grow 2.8 Percent in 2015," Gartner, March 19, 2015. [Online]. Available: http://www.gartner.com/newsroom/id/3010017 [Accessed April 15, 2015].
3. Multiple, "List of Java Virtual Machines," Wikipedia, May 17, 2014. [Online]. Available: http://en.wikipedia.org/wiki/List_of_Java_virtual_machines [Accessed June 3, 2015].
4. S. Shmeltzer, "Oracle Brings Java to iOS Devices (And Android Too)," Oracle, December 4, 2012. [Online]. Available: https://blogs.oracle.com/mobile/entry/oracle_brings_java_to_ios [Accessed June 4, 2014].
5. S. Shmeltzer, "Oracle ADF Mobile—Develop iOS and Android Mobile Applications with Oracle ADF," Oracle, October 22, 2012. [Online]. Available: https://blogs.oracle.com/jdeveloperpm/entry/oracle_adf_mobile_develop_ios [Accessed August 21, 2014].
6. Multiple, "Adobe Flash," Wikipedia, May 28, 2014. [Online]. Available: http://en.wikipedia.org/wiki/Adobe_Flash [Accessed June 3, 2014].
7. D. Meyer, "Mozilla Warns of Flash and Silverlight 'Agenda'," ZDNet, April 30, 2008. [Online]. Available: http://www.zdnet.com/news/mozilla-warns-of-flash-and-silverlight-agenda/199508 [Accessed June 4, 2014].
8. D. Winokur, "Flash to Focus on PC Browsing and Mobile Apps: Adobe to More Aggressively Contribute to HTML5," Adobe, November 9, 2011. [Online]. Available: http://blogs.adobe.com/conversations/2011/11/flash-focus.html [Accessed June 4, 2014].
9. Microsoft, "Microsoft Presents Smart Personal Objects Technology (SPOT)-Based Wristwatches at CES," Microsoft, January 9, 2003. [Online]. Available: http://www.microsoft.com/en-us/news/press/2003/jan03/01-09spotwatchespr.aspx [Accessed September 1, 2014].
10. A. J. Bandodkar, W. Jia, C. Yardımcı, X. Wang, J. Ramirez, and J. Wang, "Tattoo-Based Noninvasive Glucose Monitoring: A Proof-of-Concept Study," *Analytical Chemistry Society*, vol. 87, no. 1, pp. 394–398, 2015.
11. L. Clark, "Big Brother Is a Vending Machine: Coke Reveals Smart Fridge 2; Even Smarter," *AdNews*, May 1, 2015. [Online]. Available: http://www.adnews.com.au/news/big-brother-is-a-vending-machine-coke-reveals-smart-fridge-2-even-smarter [Accessed April 15, 2015].
12. Y. A. Pathak, "Where Is the Energy Spent Inside My App? Fine Grained Energy Accounting on Smartphones with Eprof," in Proceedings of the 7th ACM European Conference on Computer Systems, Bern, Switzerland, 2012.
13. Microsoft, "Supporting 32-Bit I/O in Your 64-Bit Driver," Microsoft, January 1, 2014. [Online]. Available: https://msdn.microsoft.com/en-us/library/windows/hardware/ff563897(v=vs.85).aspx [Accessed February 28, 2015].
14. Wikipedia, "Mutual Authentication," Wikipedia, October 29, 2014. [Online]. Available: http://en.wikipedia.org/wiki/Mutual_authentication [Accessed February 28, 2015].
15. D. E. Comer, *Internetworking with TCP/IP Principles, Protocols, and Architectures*, 4th ed., Upper Saddle River, NJ: Prentice-Hall, 2000.

Designing Platform Independent Mobile Apps and Services, First Edition. Rocky Heckman.
© 2016 the IEEE Computer Society, Inc. Published 2016 by John Wiley & Sons, Inc.

16. IETF, "HTTP—Hypertext Transfer Protocol," W3C, June 2014. [Online]. Available: http://www.w3.org/Protocols/ [Accessed February 28, 2015].

17. IETF, "Hypertext Transfer Protocol (HTTP/1.1): Message Syntax and Routing," W3C, June 2014. [Online]. Available: http://tools.ietf.org/html/rfc7230 [Accessed February 28, 2015].

18. R. T. Fielding, "Architectural Styles and the Design of Network-Based Software Architectures," Doctoral dissertation, University of California, Irvine, 2000.

19. D. Ferguson, "Exclusive .NET Developer's Journal "Indigo" Interview with Microsoft's Don Box," .NET Developer's Journal, August 10, 2004. [Online]. Available: http://dotnet.sys-con.com/node/45908 [Accessed March 1, 2015].

20. X. P. W. Group, "SOAP Version 1.2 Part 0: Primer (Second Edition)," W3C, April 27, 2007. [Online]. Available: http://www.w3.org/TR/2007/REC-soap12-part0-20070427/ [Accessed March 1, 2015].

21. R. Jandl, R. Jeyaraman, and S. Hagen, "OData Version 4.0 Part 1: Protocol Plus Errata 02," OASIS, October 30, 2014. [Online]. Available: http://docs.oasis-open.org/odata/odata/v4.0/odata-v4.0-part1-protocol.html [Accessed March 1, 2015].

22. Microsoft, "Versioning Best Practices," Microsoft, February 26, 2015. [Online]. Available: https://msdn.microsoft.com/en-us/library/azure/dn744251.aspx [Accessed March 18, 2015].

23. ECMA International, "ECMA-404: The JSON Data Interchange Format," October 2013. [Online]. Available: http://www.ecma-international.org/publications/files/ECMA-ST/ECMA-404.pdf [Accessed March 3, 2015].

24. W3C, "Extensible Markup Language (XML), W3C Working Draft 14-Nov-96," November 14, 1996. [Online]. Available: http://www.w3.org/TR/WD-xml-961114.html [Accessed March 3, 2015].

25. N. Walsh, "Deprecating XML," November 17, 2010. [Online]. Available: http://norman.walsh.name/2010/11/17/deprecatingXML [Accessed March 3, 2015].

26. C. J. Augeri, B. E. Mullins, D. A. Bulutoglu, and R. O. Baldwin, "An Analysis of XML Compression Efficiency," in Workshop on Experimental Computer Science, San Diego, 2007.

27. C. McAnlis, "The Workbench—JSON Compression : Transpose & Binary," August 14, 2013. [Online]. Available: http://mainroach.blogspot.com.au/2013/08/json-compression-transpose-binary.html [Accessed March 4, 2015].

28. W3C, "Mobile Web Application Best Practices," W3C, December 14, 2010. [Online]. Available: http://www.w3.org/TR/mwabp/}bp-presentation [Accessed March 3, 2015].

29. W3C, "Device Description Repository Simple API," W3C, December 5, 2008. [Online]. Available: http://www.w3.org/TR/DDR-Simple-API/ [Accessed March 3, 2015].

30. D. McMurty, A. Oakely, M. Subramanian, H. Zhang, and J. Sharp, *Data Access for Highly-Scalable Solutions: Using SQL, NoSQL, and Polyglot Persistence*, Redmond: Microsoft, 2013.

31. Wikipedia, "CAP Theorem," Wikipedia, January 15, 2015. [Online]. Available: http://en.wikipedia.org/wiki/CAP_theorem [Accessed February 19, 2015].

32. E. Brewer, "CAP Twelve Years Later: How the "Rules" Have Changed," *IEEE Explore*, vol. 45, no. 2, pp. 23–29, 2012.

33. A. Homer, J. Sharp, L. Brader, M. Narumoto, and T. Swanson, *Cloud Design Patterns*, Redmond: Microsoft, 2014.

34. G. Hohpe and B. Woolf, *Enterprise Integration Patterns: Designing, Building, and Deploying Messaging Solutions*, Boston: Addison-Wesley Professional, 2003.

35. M. Fowler, *Patterns of Enterprise Application Architecture*, Boston: Addison-Wesley Professional, 2003.

36. Unity 3D, "Unity Public Relations," Unity 3D, February 19, 2015. [Online]. Available: http://unity3d.com/public-relations [Accessed February 19, 2015].

37. Microsoft, "Microsoft Takes .NET Open Source and Cross-Platform, Adds New Development Capabilities with Visual Studio 2015, .NET 2015 and Visual Studio Online," Microsoft, November 12, 2014. [Online]. Available: http://news.microsoft.com/2014/11/12/microsoft-takes-net-open-source-and-cross-platform-adds-new-development-capabilities-with-visual-studio-2015-net-2015-and-visual-studio-online/ [Accessed March 3, 2015].

38. K. Brockschmidt and M. Jones, "Cross-Platform: Write Cross-Platform Hybrid Apps in Visual Studio with Apache Cordova," *msdn Magazine*, pp. 4–13, December 2014.

39. IBM, "WebSphere MQ Queue Properties," IBM, October 3, 2014. [Online]. Available: http://www-01.ibm.com/support/knowledgecenter/SSFKSJ_8.0.0/com.ibm.mq.explorer.doc/e_properties_queues.htm [Accessed November 8, 2014].

40. Amazon, "Amazon Simple Queue Service," Amazon, January 15, 2015. [Online]. Available: http://aws.amazon.com/sqs/ [Accessed January 15, 2015].

41. Microsoft, "Service Bus Pricing and Billing," Microsoft, February 3, 2015. [Online]. Available: https://msdn.microsoft.com/en-us/library/dn831889.aspx [Accessed April 14, 2015].

42. Microsoft, "Microsoft Azure Service Bus," Microsoft, October 27, 2014. [Online]. Available: http://msdn.microsoft.com/en-us/library/azure/ee73257.aspx [Accessed November 10, 2014].

43. S. Manheim, "How to Use the Service Bus Relay Service," Microsoft, February 18, 2015. [Online]. Available: http://azure.microsoft.com/en-us/documentation/articles/service-bus-dotnet-how-to-use-relay/ [Accessed April 14, 2015].

44. IETF, "RFC 6455," IETF, December 1, 2011. [Online]. Available: http://tools.ietf.org/html/rfc6455 [Accessed November 10, 2014].

45. W. Vogels, "Making Mobile App Development Easier with Cross Platform Mobile Push," Amazon, August 13, 2013. [Online]. Available: http://www.allthingsdistributed.com/2013/08/amazon-sns-mobile-push.html [Accessed February 20, 2015].

46. Windows App Team, "Delivering Push Notifications to Millions of Devices with Windows Azure Notification Hubs," Microsoft, September 16, 2013. [Online]. Available: http://blogs.windows.com/buildingapps/2013/09/16/delivering-push-notifications-to-millions-of-devices-with-windows-azure-notification-hubs/ [Accessed February 20, 2015].

47. Microsoft, "Sochi Olympics Site Uses the Cloud to Support More Than 90,000 Requests Per Second," Microsoft, September 17, 2014. [Online]. Available: http://customers.microsoft.com/Pages/CustomerStory.aspx?recid=7942 [Accessed February 20, 2015].

48. Technical Committee ISO/TC 211, Geographic information/Geomatics, "Address Standards, ISO 19160, Addressing," ISO, December 8, 2011. [Online]. Available: http://www.isotc211.org/address/iso19160.htm [Accessed February 21, 2015].

49. Universal Postal Union, "Addressing the World—An Address for Everyone," Universal Postal Union, January 2010. [Online]. Available: http://www.upu.int/fileadmin/documentsFiles/activities/addressingAssistance/paperAddressingAddressingTheWorldAnAddressForEveryoneEn.pdf [Accessed February 21, 2015].

50. W3C, "Web Services Description Language (WSDL) Version 2.0 Part 1: Core Language," W3C, June 26, 2007. [Online]. Available: http://www.w3.org/TR/wsdl20/ [Accessed February 21, 2015].

51. B. Meyer, *Object-Oriented Software Construction*, 2nd ed., Prentice Hall, 1997.

52. E. Evans, *Domain-Driven Design: Tackling Complexity in the Heart of Software*, Addison-Wesley, 2003.

53. C. Munoz, "Facebook Architecture Presentation: Scalability Challenge," July 12, 2014. [Online]. Available: http://www.slideshare.net/Mugar1988/facebook-architecture-presentation-scalability-challenge [Accessed March 20, 2015].

54. H. Dierking, "Versioning RESTful Services," CodeBetter.com, November 9, 2012. [Online]. Available: http://codebetter.com/howarddierking/2012/11/09/versioning-restful-services/ [Accessed March 18, 2015].

55. Microsoft, "Entity Framework Class Library 6.0," Microsoft, 2014. [Online]. Available: https://msdn.microsoft.com/en-us/library/dn223258(v=vs.113).aspx [Accessed March 10, 2015].

56. Multiple, "Entity Framework," Wikipedia, December 31, 2014. [Online]. Available: http://en.wikipedia.org/wiki/Entity_Framework [Accessed March 10, 2015].

57. D. G. Low (interviewee), "NoSql versus Relational Databases," [Interview], March 10, 2015.

58. R. Miller, "Codeplex EF 5.0.0," Microsoft, October 28, 2013. [Online]. Available: https://entityframework.codeplex.com/releases/view/86738 [Accessed March 10, 2015].

59. M. Cline, "Serialization and Unserialization," C++ Foundation, 2015. [Online]. Available: https://isocpp.org/wiki/faq/serialization [Accessed April 5, 2015].

60. JSON.org, "Introducing JSON," JSON.org, December 1999. [Online]. Available: http://json.org [Accessed April 5, 2015].

61. Wikipedia, "Database," March 17, 2015. [Online]. Available: http://en.wikipedia.org/wiki/Database [Accessed March 23, 2015].

62. E. F. Codd, "Recent Investigations into Relational Data Base Systems," IBM Research, New York, 1974.

63. PeaZip, "Compression/Extraction Benchmarks of WinZip, WinRar, PeaZip," PeaZip, 2013. [Online]. Available: http://peazip.sourceforge.net/peazip-compression-benchmark.html [Accessed March 10, 2015].

64. Apple, "Replacing Your App with a New Version," Apple, April 9, 2015. [Online]. Available: https://developer.apple.com/library/ios/documentation/LanguagesUtilities/Conceptual/iTunesConnect_Guide/Chapters/ReplacingYourAppWithANewVersion.html#//apple_ref/doc/uid/TP40011225-CH14 [Accessed April 11, 2015].

65. Google, "Update Your Apps," Google, 2015. [Online]. Available: https://support.google.com/googleplay/android-developer/answer/113476?hl=en [Accessed April 11, 2015].

66. Microsoft, "Submitting and App Update," Microsoft, 2015. [Online]. Available: https://developer.microsoft.com/en-us/windows/publish [Accessed April 11, 2015].

67. N. Lovell, "Conversion Rate," Gamesbrief, November 9, 2011. [Online]. Available: http://www.gamesbrief.com/2011/11/conversion-rate/ [Accessed April 11, 2015].

68. Zynga Inc., "Zynga Q2 2014 Report Filing," Zynga Inc., San Francisco, 2014.

69. Wikipedia, "Chinglish," April 7, 2015. [Online]. Available: http://en.wikipedia.org/wiki/Chinglish [Accessed April 11, 2015].

70. Fraunhofer, "Research News: Measuring Glucose with Needle Pricks," Fraunhofer, Munich, 2012.

INDEX

Active Server Pages (ASP), 5
Adobe AIR, 6
Adobe Flash Player, 6
Aggregates, 131
Amazon Device Messaging, 91
Amazon's Simple Queue Service, 86
Android, 1
Apache Cordova, 7
Apple, 1
Apple Push Notification Service, 91
Arduino, 13
Asynchronous CE, 101
Azul Zing JVM, 3

Baidu Cloud Push for Android, 91
Blackberry, 1
bounded contexts, 131
Boyce-Codd Normal Form, 190
breadth versus depth, 9

CAP Theorem, 55
CIDR, 23
Claim Check pattern, 102
client-server, 11
Code First, 155
Column Family Databases, 189
   Super Column, 193
Command Query Responsibility
      Separation, 127
Compensating Transaction pattern, 57
compensation logic, 57
concurrency, 139
content filter, 102
CQRS, 127
   CQRS/ES, 127
   read side, 132
   write side, 132

Data Abstraction Layer, 70, 154
Data Layer, 70, 176
Data Mapper pattern, 154
data transfer object, 136
document database, 186
DocumentDB, 33
Domain Driven Design, 128
DynamoDB, 33

ECMAScript, 6
embedded, 13
Entity Framework, 155
Event Hubs, 86
Event Sourcing, 134
Eventual Consistency, 131

file storage, 197
Flash, 6
Free and Open Source Software, 17

Google, 1
Google Cloud Messaging, 91
Graph Database, 194

HADR, 151
HATEOAS, 26
HoloLens, 11
horizontal scaling, 115
HTML 5, 6
HTTP, 20

IBM J9, 3
Idempotent Receiver pattern, 58
Internet of Things, 11
iPad, 1
iPhone, 1
isolated storage, 9